HACCP

Principles and Applications

Edited by

Merle D. Pierson

and

Donald A. Corlett, Jr.

*Based on a short course developed and presented
by the Continuing Education Committee of the
Institute of Food Technologists*

CHAPMAN & HALL
New York • London

Copyright © 1992 by Van Nostrand Reinhold
Softcover reprint of the hardcover 1st edition 1992

ISBN 978-1-4684-8820-3   ISBN 978-1-4684-8818-0 (eBook)
DOI 10.1007/978-1-4684-8818-0

Library of Congress Catalog Card Number 92-6227

This edition published by
Chapman & Hall
One Penn Plaza
New York, NY 10119

Published in Great Britain by
Chapman & Hall
2-6 Boundary Row
London SE1 8HN

16 15 14 13 12 11 10 9 8 7 6 5 4

Library of Congress Cataloging-in-Publication Data

HACCP : principles and applications / edited by Merle D. Pierson and
   Donald A. Corlett, Jr.
      p.   cm.
   "Based on a short course developed and presented by the Continuing
Education Committee of the Institute of Food Technologists."
   "An AVI book."
   Includes bibliographical references and index.

   1. Food—Microbiology.   2. Food—Safety measures.   3. Food
industry and trade—Quality control.   I. Pierson, Merle D.
II. Corlett, Donald A.   III. Institute of Food Technologists.
Continuing Education Committee.
QR115.H23   1992
664'.07--dc20   92-6227
                                                         CIP

British Library Cataloguing in Publication Data available

Please send your order for this or any **Chapman & Hall book to Chapman & Hall, 29 West 35th Street, New York, NY 10001, Attn: Customer Service Department.** You may also call our Order Department at 1-212-244-3336 or fax your purchase order to 1-800-248-4724.

For a complete listing of Chapman & Hall's titles, send your requests to **Chapman & Hall, Dept. BC, One Penn Plaza, New York, NY 10119.**

Contents

7 Monitoring Critical Control Point Critical Limits
Martha Hudak-Roos and E. Spencer Garrett   62

8 Corrective Action Procedures for Deviations from the Critical Control Point Critical Limits *R. B. Tompkin*   72

9 Effective Recordkeeping System for Documenting the HACCP Plan *K. E. Stevenson and Bonnie J. Humm*   83

Preface

The Institute of Food Technologists (IFT) sponsors each year a two-day short course that covers a topic of major importance to the food industry. "Hazard Analysis and Critical Control Points" was the title for the short course which was held May 31–June 1, 1991, immediately prior to the 51st Annual IFT Meeting. These short courses have been published as a proceedings in previous years; however, the current and future importance of the Hazard Analysis and Critical Control Point (HACCP) system prompted publication of the 1991 short course as a book. This book is designed to serve as a reference on the principles and application of HACCP for those in quality control/assurance, technical management, education and related areas who are responsible for food safety management.

The National Advisory Committee on Microbiological Criteria for Foods (NACMCF) published in November 1989 a pamphlet titled "HACCP Principles for Food Production" (Appendix A). This document dealt with HACCP as applied to the microbiological safety of foods; however, the principles can be modified to apply to chemical, physical and other hazards in foods. The principles recommended by the NACMCF have been widely recognized and adopted by the food industry and regulatory agencies. Implementation of these principles provides a proactive, preventive system for managing food safety. HACCP should be applied at all stages of the food system, from production to consumption. In order for HACCP to be effective, however, specific plans must be designed for each stage of the food system, specific food products, different food operations, etc.

This book is based on the seven principles recommended by the NACMCF. However, the topic has been broadened to include all biological, chemical, and physical hazards. The first three chapters provide an introduction to HACCP,

HACCP principles and definitions, and an overview of biological, chemical, and physical hazards in foods. The next seven chapters give an in-depth discussion and analysis of each of the HACCP principles. HACCP as presented by the NACMCF relates to food safety concerns only; however, some government agencies and companies use control points other than those relating to safety within the context of HACCP. Therefore, a chapter is included on safety, quality and regulatory control points. Next, an action plan for implementing HACCP is given as well as an overview of the adoption of HACCP by federal regulatory agencies. Several specific examples for the application of HACCP principles are provided in the last chapter.

The general concept of HACCP as applied to food safety management has not changed since it was first publicly introduced at the 1971 Conference on Food Protection. Practical experience and a wider application of HACCP over the years has led to revisions in principles, definitions, and methods of application. This book, as explained above, is based on the November 1989 principles recommended by the NACMCF. In June, 1991 the Codex Committee on Food Hygiene HACCP Drafting Group developed a *draft* report on HACCP. This draft report is given in Appendix B. The NACMCF proposed in July, 1991, revision of the HACCP principles and their application. The revised HACCP document that was approved on March 20, 1992 by the NACMCF is given in Appendix C. The principles of HACCP as provided by the NACMCF gives an excellent framework and common approach for industry and regulatory to manage food safety. At the same time, it needs to be pointed out that the principles serve as guidelines in developing HACCP plans for food products and the application of these principles may need to be modified based on advances in technology, practical experience, etc. The discussion presented in this book is based on the NACMCF November 1989 principles document Appendix A. This information will continue to serve as a valuable reference when adopting revisions of the HACCP principles.

We thank Dr. Robert Price, Chairman, and the members of the IFT Short Course Committee for their assistance and suggestions in developing the 1991 HACCP Short Course. Appreciation is extended to Dr. Daryl Lund, 1990–91 IFT President; Howard Mattson, former IFT Executive Director; John B. Klis, IFT Director of Publications; and members of the IFT staff who provided publicity, facility planning, and numerous other details for a successful meeting. Special recognition is given to Anna May Schenck for her skillful copy editing of the manuscripts. Finally, we are grateful to the contributing authors for taking time from their demanding schedules to prepare manuscripts for this book.

Merle D. Pierson
Donald A. Corlett, Jr.

Contributors

Howard E. Bauman
1433 Utica Ave. S.
Suite 70-4
St. Louis Park, MN 55416

Dr. Donald A. Corlett
Corlett Food Consulting Services
5745 Amaranth Pl.
Concord, CA 94521

Merle D. Pierson
Department of Food Science and Technology
VPI and State University
Blacksburg, VA 24061

E. Jeffery Rhodehamel
Division of Microbiology
Food and Drug Administration
Washington, DC 20204

William H. Sperber
Grand Metropolitan Food Sector
330 University Ave. SE
Minneapolis, MN 55414

Lloyd J. Moberg
General Mills, Inc.
No. 1 General Mills Blvd.
Minneapolis, MN 55426

Martha Hudak-Roos
National Seafood Inspection Laboratory
P.O. Drawer 1207
Pascagoula, MS 39567

E. Spencer Garrett
National Seafood Inspection Laboratory
P.O. Drawer 1207
Pascagoula, MS 39567

R. B. Tompkin
Armour Swift-Eckrick, Inc.
3131 Woodcreek Dr.
Downers Grove, IL 60515

K.E. Stevenson
Microbiology/Sanitation Division
Western Research Laboratory
National Food Processors Association
6363 Clark Ave.
Dublin, CA 94568

Bonnie J. Humm
Microbiology/Sanitation Division
Western Research Laboratory
National Food Processors Association
6363 Clark Ave.
Dublin, CA 94568

Gale Prince
The Kroger Company
1014 Vine St.
Cincinnati, OH 45201

John Humber
Microbiology and Food Safety
Kraft General Foods
801 Waukegan Rd.
Glenview, IL 60025

Catherine E. Adams
Grocery Manufacturers of America, Inc.
1010 Wisconsin Ave., NW
Washington, DC 20007

Dale D. Boyle
Naval Supply Systems Command
Crystal Mall, Bldg. 3, Room 710
1931 Jefferson Davis Highway
Arlington, VA 20226

Richard Stier
Libra Laborfatories, Inc.
44 Stelton Rd.
Piscataway, NJ 08854

1

Introduction to HACCP

Howard E. Bauman

The concept and reduction to practice of the Hazard Analysis and Critical Control Point (HACCP) system was directly related to the Pillsbury Company's projects in food production and research for the space program. The basics were developed by the Pillsbury Company with the cooperation and participation of The National Aeronautics and Space Agency (NASA), the Natick Laboratories of the U.S. Army, and the U.S. Air Force Space Laboratory Project Group.

The pathway to the HACCP system started in 1959 when Pillsbury was asked to produce a food that could be used under zero gravity conditions in the space capsules. We started with the fact that no one really knew how foods, especially particulates, might act in zero gravity. The initial conservative approach to solve this problem was to produce bite-sized foods covered with a flexible edible coating to prevent crumbling and consequently atmospheric contamination. The most difficult part of the program, however, was to come as close to 100% assurance as possible that the food products we were producing for space use would not be contaminated with bacterial or viral pathogens, toxins, chemicals, or physical hazards that could cause an illness or injury. Such hazards might result in an aborted or catastrophic mission.

It was quickly determined that by using existing techniques of quality control there was no way we could be assured that there wouldn't be a problem. Further, the amount of testing that had to be done to arrive at a reasonable decision point as to whether a food was acceptable was extremely high. In fact, a large part of the production of any particular batch of food had to be utilized for testing, leaving only a small portion available for the space flights.

This raised two questions. First, "What could we do using new techniques that would help us approach the 100% assurance level?" Second, since companies for good reason didn't practice this type of destructive testing, "How much in the way of hazards was the industry missing by minimal tests of the raw materials, and in-line and end product tests?"

This brought into serious question the prevailing system of quality control that was being used in our plants and the food industry as a whole. Most quality assurance programs were based on what the quality assurance manager believed was a good program. There was no uniformity of approach or even understanding in the food industry as to what constituted an excellent program. In our search for answers, we examined the zero defects program utilized by NASA and found that it was designed for hardware. The type of testing that was used for hardware, such as x-ray and ultrasound was nondestructive and, therefore, suitable for this purpose but not for food.

In looking for a better system, we decided to try a new approach to the problem. We concluded after extensive evaluation that the only way we could succeed would be to develop a preventive system. This would require us to have control over the raw materials, process, environment, personnel, storage, and distribution as early in the system as we possibly could. We felt certain that if we could establish this type of control, along with appropriate record keeping, we should be able to produce a high degree of assurance a product we could say was safe. For all practical purposes if this system was implemented correctly, there would be no testing of the finished packaged product other than for monitoring purposes. It should also be noted that the type of record keeping required under NASA rules not only furnished a clue as to how to approach the new system, but also facilitated our experimentation with this approach and is a basic part of the HACCP system as it now exists.

We were required by NASA to keep records that allowed traceability of the raw materials we used, the plant where the food was produced, the names of people involved in the production and any other information that might contribute to the history of the product. In other words, a mechanism for tracing problems back to the source. This required the development of a familiarity with the raw materials that was not a normal process in food product development. For instance, we knew the latitude and longitude where the salmon used in salmon loaf were caught as well as the name of the ship. It was using this approach that we developed the HAACP system.

HACCP is a preventive system of quality control. The system when properly applied can be used to control any area or point in the food system that could contribute to a hazardous situation whether it be contaminants, pathogenic microorganisms, physical objects, chemicals, raw materials, a process, use directions for the consumer or storage conditions.

The Hazard Analysis portion of HACCP involves a systematic study of the ingredients, the food product, the conditions of processing, handling, storage, packaging, distribution and consumer use. This analysis allowed us to identify in the process flow the sensitive areas that might contribute to a hazard. From this information we were able to determine the Critical Control Points in the system that had to be monitored.

The definition of a Critical Control Point is any point in the chain of food production from raw materials to finished product where the loss of control could result in an unacceptable food safety risk.

Our first problem using this approach was that we knew what we wanted to do, but didn't know how to do an adequate hazard analysis. While searching for a method, we found that the U.S. Army Natick Research, Development, and Engineering Center had developed a system of analysis called modes of failure which was used for medical supplies. After evaluating this method, we adopted the modes of failure technique with some modifications as our model. We also found that to do an adequate analysis we had to break down each ingredient and product and its production system into its components and analyze each segment for its potential contribution to safety. When this was completed, it was necessary to connect them all together to develop the overall interrelationship. This is critical, because whenever changes are made in an approved interrelated system, the system must be reevaluated since any change in the system—even though it may appear innocuous—could have a major effect downstream in the system.

We approached the problem by starting with the raw materials. We looked at specific ingredients as well as each stage of processing from the field through the food chain. This was done to determine what might happen to raw materials and what we might expect in the way of problems when they appeared at the plant. It was from these analyses, including searches of the literature, discussions with suppliers and of course our own history of the ingredients that we were able to select those sensitive ingredients and sensitive areas that must be monitored and controlled in order to insure that we would not bring a hazard into the plant.

The areas of concern ranged from the potential presence of pathogens, heavy metals, toxins, physical hazards and chemicals, to the type of treatments the ingredients might have received such as pesticide applications or a pasteurization step. The next segment was an analysis of the manufacturing process, the building, the general environment and method of people control. This was done to ensure that we completely understood all of the points or areas in the facilities and process that might contribute to a hazard. It also included determining those procedures that would prevent a hazard. Another segment of investigation was the examination of the storage, transportation and distribution to be used for the product and the abuses it might receive. Finally an analysis was conducted to determine what the consumer might do to the product that could cause unsafe conditions.

This is a rather simplified sketch of what must be done. However, it does show that detailed knowledge of the total system for the production of any food must be developed.

The HACCP system was first exposed to the public during the 1971 National Conference on Food Protection (U.S. Dept. HEW 1972). Following this conference, Pillsbury was granted a contract by FDA to conduct classes for FDA personnel on the HACCP system. The first comprehensive document on HACCP was published by the Pillsbury Company (1973) and was used for training FDA inspectors in HACCP principles. A special session was held with personnel involved in FDA's acidified and low-acid canned food regulation. This group developed the necessary information for the promulgation of the acidified and low-acid canned food regulation (FDA 1973) which is a successful HACCP system.

HACCP has been used in the plants of the Pillsbury Company since 1971. During the 1970's and early 1980's a number of companies requested and were given information and help in establishing their own HACCP programs. It wasn't until 1985 that the HACCP system was seriously considered for broad application in the food industry. In 1985 the HACCP system was recommended by the National Academy of Sciences (NAS) in the publication *An Evaluation of the Role of Microbiological Criteria for Foods and Food Ingredients* (NAS 1985). The NAS Committee (Subcommittee on Microbiological Criteria for Foods and Food Ingredients) concluded that a preventive system (HACCP) was essential for control of microbiological hazards. They concluded that end product testing was not adequate to prevent foodborne disease. In 1987, the National Oceanic and Atmospheric Administration (NOAA) was charged by congress "to design a program of certification and surveillance to improve the inspection of fish and seafood consistent with the hazard analysis critical control points system." This effort has been carried out by the National Marine Fisheries Service (see Chapter 13).

The National Academy of Sciences publication also recommended that a National Advisory Committee on Microbiological Criteria for Foods be established. This has been accomplished and the committee has been active not only in developing microbiological criteria, but has embraced the HACCP concept. The committee has further refined HACCP by adding to the principles of HACCP appropriate descriptions of what each principle involves. They have also developed definitions of terminology used in HAACP. The HACCP document of this committee is intended to be a guide for maintaining a uniform system through the use of the principles and definitions. This approach will make possible a universal system of food safety that should facilitate the movement of foods internationally as well as providing a high level of assurance that the foods are safe.

There will undoubtedly be refinements in the future as more experience with the system is gained. It is easy to modify HACCP systems, however, they should be done with care and general agreement that the modifications add to the reliability of the system and will not degrade it.

References

National Academy of Sciences (NAS). 1985. *An Evaluation of the Role of Microbiological Criteria for Foods and Food Ingredients*. NAS, National Research Council, National Academy Press, Washington, DC.

Pillsbury Company. 1973. Food Safety Through the Hazard Analysis and Critical Control Point System. Contract No. FDA 72-59. Research and Development Dept., The Pillsbury Company, Minneapolis, MN 55414.

U.S. Dept. of Health, Education and Welfare. 1972. *Proceedings of the 1971 Conference on Food Protection*. U.S. Government Printing Office, Washington, DC.

U.S. Food and Drug Administration. 1973. Acidified Foods and Low-Acid Foods in Hermetically Sealed Containers. Code of U.S. Federal Regulations, Title 21, Ch. 1, Parts 113 and 114 (renumbered since 1973).

Chapter 2

HACCP: Definitions and Principles

Donald A. Corlett, Jr.
Merle D. Pierson

INTRODUCTION

HACCP definitions and principles covered in this book are based on the U.S. National Advisory Committee on Microbiological Criteria for Foods (NACMCF) HACCP system guide, approved November, 1989. The NACMCF guide is published in the pamphlet, "HACCP Principles for Food Production," (NACMCF 1989) which is given in Appendix A. This document will serve as the basis for HACCP definitions and principles for this volume.

The NACMCF guide defines HACCP as a systematic approach to be used in food production as a means to assure food safety. Definitions, principles, a description of each principle, and the implementation guide are intended to provide a clear and useful format for individual food industry producers to develop their own HACCP systems tailored to their specific products, processing, and distribution conditions.

This chapter reviews the definitions and principles, as well as the implementation guide contained in the pamphlet. For convenience, the table of contents of "HACCP Principles for Food Production," (Appendix A) is given as follows:

6

It should be noted that the NACMCF document only covers microbiological hazard analysis under Principle 1. "Assess hazards associated with growing, harvesting, raw materials and ingredients, processing, manufacturing, distribution, marketing, preparation and consumption of the food." However, the Committee intended the HACCP principles to eventually cover chemical and physical hazards in food.

Recently a protocol was developed by Corlett and Stier (1991) that extended the system to chemical and physical hazards in food. This was patterned after the six microbiological hazard characteristics described in the pamphlet under Principle 1. Complete microbiological, chemical and physical hazard analysis procedures are described in Chapter 4, "Hazard Analysis and Assignment of Risk Categories." The combined microbiological, chemical and physical hazard risk assessment permits application of HACCP for all classes of food hazards.

The NACMCF HACCP definitions and principles were transmitted and recommended in 1989 to four federal departments: Agriculture, Health and Human Services, Commerce, and Defense. Approval of the guide by the NACMCF culminated an intensive eight-month project by the *ad hoc* HACCP working group composed of Dr. Catherine E. Adams, USDA-FSIS; Dr. Howard Baumen, Pillsbury Company (ret.); chairman Dr. Donald A. Corlett Jr., ESCAgenetics Corporation; Mr. Cleve B. Denny, National Food Processors Association; Mr. Spencer Garrett, National Marine Fisheries Service; Dr. John Kvenberg, Food and Drug Administration; Dr. David M. Theno, Foster Farms; and Dr. Bruce Tompkin, Swift-Eckrich, Inc.

References

Corlett, D.A. and Stier, R.F. 1991. Risk assessment within the HACCP system. Food control. 2:71–72.

National Advisory Committee on Microbiological Criteria for Foods (NACMCF). 1989. *HACCP Principles for Food Production.* USDA-FSIS Information Office, Washington, DC 20250.

3

Overview of Biological, Chemical, and Physical Hazards

E. Jeffery Rhodehamel

INTRODUCTION

HACCP is a systematic approach to be used in food production as a means to ensure food safety. The first step requires a hazard analysis, an assessment of risks associated with all aspects of food production from growing to consumption. However, before one can assess the risks, a working knowledge of potential hazards must be obtained. A hazard is defined by the National Advisory Committee on Microbiological Criteria for Foods (NACMCF) as any biological, chemical, or physical property that may cause an unacceptable consumer health risk. Thus, by definition one must be concerned with three classes of hazards; biological, chemical, and physical.

This chapter provides a generalized background of the potential hazards associated with foods. Appropriate reference materials on food hazards have been included. A number of textbooks are available on the subject of hazards in foods.

BIOLOGICAL HAZARDS

The first hazard category, biological or microbiological, can be further divided into three types: bacterial, viral, and parasitic (protozoa and worms). Many HACCP programs are designed specifically around the microbiological hazards. Archer and Kvenberg (1985) and Todd (1989) estimated that the incidence of foodborne illness ranges from 12.6 to 81 million cases per year with a cost of 1.9 to 8.4 billion dollars. HACCP programs address this food safety problem by assisting in the production of safe wholesome foods. Excellent references exist on biological hazards (Cliver 1990), foodborne pathogenic bacteria (Doyle

8

1989; Riemann and Bryan 1979), viruses (Cliver 1988), and parasitic protozoa and worms (Healy et al. 1984; Jackson 1990).

Table 3-1 lists hazardous bacteria, viruses, and parasitic protozoa and worms, which include the microorganisms of concern in HACCP programs. The International Commission of Microbiological Specifications for Food (ICMSF 1986) attempted to group some of these hazardous microorganisms according to severity of risk (Table 1). The pathogens in Group I present a severe hazard; those in

TABLE 3-1   Hazardous Microorganisms and
Parasites Grouped on the Basis of Risk
Severity[a]

I. Severe Hazards
    Clostridium botulinum types A, B, E, and F
    Shigella dysenteriae
    Salmonella typhi; paratyphi A, B
    Hepatitis A and E
    Brucella abortis; B. suis
    Vibrio cholerae O1
    Vibrio vulnificus
    Taenia solium
    Trichinella spiralis
II. Moderate Hazards: Potentially Extensive Spread[b]
    Listeria monocytogenes
    Salmonella spp.
    Shigella spp.
    Enterovirulent *Escherichia coli* (EEC)
    Streptococcus pyogenes
    Rotavirus
    Norwalk virus group
    Entamoeba histolytica
    Diphyllobothrium latum
    Ascaris lumbricoides
    Cryptosporidium parvum
III. Moderate Hazards: Limited Spread
    Bacillus cereus
    Campylobacter jejuni
    Clostridium perfringens
    Staphylococcus aureus
    Vibrio cholerae, non-O1
    Vibrio parahaemolyticus
    Yersinia enterocolitica
    Giardia lamblia
    Taenia saginata

---

[a]Adapted from ICMSF (1986).
[b]Although classified as moderate hazards, complications and sequelae may be severe in certain susceptible populations.

Group II are considered moderate hazards (although the illnesses in certain susceptible populations or complications can be severe) with the potential for extensive spread of the disease. Pathogens in Group III cause common-source outbreaks; however, subsequent spread is either rare or limited.

When developing a HACCP program, the food grower or processor should have three basic aims with regard to biological hazards: (1) destroy, eliminate, or reduce the hazard; (2) prevent recontamination; and (3) inhibit growth and toxin production. Preventive measures should be taken to achieve these goals.

Microorganisms can be destroyed or eliminated by thermal processing, freezing, and drying. After the microorganism has been eliminated, measures to prevent recontamination should be taken. Finally, if the hazard cannot be totally eliminated from the food, microbial growth and toxin production must be inhibited. Growth can be inhibited through the intrinsic characteristics of the food, such as pH and water activity (a_w), or by the addition of salt or other preservatives. Conditions under which the food is packaged (aerobic or anaerobic) and storage temperatures (refrigeration or freezing) can also be used to inhibit growth.

Bacterial hazards

Bacterial hazards can result either in foodborne infections or intoxications. A foodborne infection is caused by ingesting a number of pathogenic microorganisms sufficient to cause infection, and the reaction of tissues to their presence, multiplication, or elaboration of toxins. A foodborne intoxication is caused by the ingestion of preformed toxins produced and excreted by certain bacteria when they multiply in foods (Bryan 1979).

Table 3-2 lists the source, disease characteristics, and food associated with various foodborne bacterial pathogens. Although not all-inclusive, the list represents the pathogens that are reported to cause foodborne disease outbreaks and that are responsible for numerous cases of illness. Many food commodities have a unique microbiology and group of associated pathogens. Processors of specific foods (e.g., seafood) should consult reference materials in those areas (e.g., Ward and Hackney 1991).

The following summary discusses the notable characteristics of the various foodborne bacterial pathogens of concern to the food industry and their relationship to the development of a HACCP program. The natural incidence and the severity of disease caused by these bacteria, along with the general conditions required for their control represent a cross-section of challenges for HACCP programs. If these organisms are controlled, numerous other pathogens may be similarly controlled.

Clostridium botulinum. *Clostridium botulinum*, the causative agent of botulism (foodborne intoxication), is an anaerobic, sporeforming rod that produces a potent

TABLE 3-2 Various Foodborne Pathogenic Microorganisms of Concern

Microorganism	Source in Nature	Characteristics of Illness	Associated Foods
Clostridium botulinum	Soil, sediment intestinal tracts of fish, mammals, gills, viscera of fish crabs, seafood	Neurotoxicity; shortness of breath, blurred vision; loss of motor capabilities; death. Onset ranges between 12 and 36 hours.	Low-acid canned foods, especially home canned. Meats, fish, smoked/fermented fish, vegetables, other marine products
Clostridum perfringens	Soil and sediment (widespread), water, intestinal tracts of humans and animals	Nausea, occasional vomiting, diarrhea and intense abdominal pain. Onset ranges from 8 to 22 hours. Short duration (24 hours).	Improperly prepared roast beef, turkey, pork, chicken, cooked ground meat and other meat dishes, gravies, soups and sauces
Salmonella spp.	Water, soil, mammals, birds, insects, intestinal tracts of animals, especially poultry and swine	Nausea, vomiting, abdominal cramps, diarrhea, fever and headache. Normal incubation period 6–48 hours.	Beef, turkey, pork, chicken eggs and products, meat salads, crabs, shellfish, chocolate, animal feeds, dried coconut, baked goods and dressings
Listeria monocytogenes	Soil, silage, water and other environmental sources, birds, mammals, and possibly fish and shellfish	Healthy individuals generally have mild flu-like symptoms. Severe forms of listeriosis include septicemia, meningitis, encephalitis, and abortion in pregnant women.	Raw milk, soft cheese, cole slaw, ice cream, raw vegetables, raw-meat sausages, raw and cooked poultry, raw and smoked fish

(continued)

TABLE 3-2 Continued

Microorganism	Source in Nature	Characteristics of Illness	Associated Foods
Campylobacter jejuni	Soil, sewage, sludge, untreated waters, intestinal tracts of chickens, turkeys, cattle, swine, rodents, and some wild birds	Fever, headache, nausea, muscle pain, and diarrhea (sometimes watery, sticky, or bloody). Onset time is 2–5 days and last 7–10 days. Relapses are common.	Raw milk, chicken, other meats and meat products
Staphylococcus aureus	Hands, throats, and nasal passages of humans; common on animal hides	Nausea, vomiting, diarrhea, abdominal cramps, and prostration. Symptoms may be severe. Normal onset is from 30 minutes to 8 hours. Duration is usually 24–48 hours.	Ham, turkey, chicken, pork, roast beef, eggs, salads (e.g., egg, chicken, potato, macaroni), bakery products, cream-filled pastries, luncheon meats, milk and dairy products
Shigella spp.	Polluted water and intestinal tracts of humans and other primates	Diarrhea with bloody stools, abdominal cramps, and fever. Severe cases caused by *S. dysenteriae* may result in septicemia, pneumonia, or peritonitis. Onset averages 0.5–2 days but may be as high as 7 days. Recovery is slow.	Milk and dairy products, raw vegetables, poultry, and salads (e.g., potato, tuna, shrimp, macaroni, and chicken)
Vibrio parahaemolyticus	Estuarine and marine waters	Abdominal cramps, nausea, vomiting, headache, and diarrhea (with occasional blood and mucus in stools) and fever. Onset ranges from 4 to 96 hours. Symptoms last an average of 2.5 days.	Raw, improperly cooked, or cooked, recontaminated fish, shellfish, or crustacea

Organism	Source	Symptoms	Foods
Vibrio cholerae O1	Untreated water, intestinal tracts of humans	Copious watery stools, vomiting, prostration, dehydration, muscular cramps, and occasionally death. Incubation ranges from 1 to 5 days.	Shell fish, raw fish, and crustacea
Bacillus cereus	Soils, sediments, dust, water, vegetation, and a variety of foods, notably cereals, dried foods, spices, milk and dairy products, meat products, and vegetables	TYPE I-Diarrheal type food poisoning—watery diarrhea, abdominal cramps, nausea, usually no fever or vomiting. Onset time 6–15 hours. Short duration (24 hours). TYPE II-Emetic type food poisoning—Nausea and vomiting within 0.5 to 6 hours. Abdominal cramps and diarrhea occasionally occur. Short duration (less than 24 hours).	Meats, vegetable dishes, milk, cream pastries, soups, and puddings — Fried, boiled; or cooked rice, and other starchy foods (e.g., potatoes and pasta)
Yersinia enterocolitica	Soil, natural waters, intestinal tracts of various animals; (pigs, birds, beavers, dogs and cats)	Diarrhea, and/or vomiting, fever and abdominal pain are hallmark symptoms; pseudoappendicitis. Onset is 24–48 hours. Recovery in 1–2 days.	Fresh meat and meat products (particularly swine), fresh vegetables, milk, and milk products
Escherichia coli (Enterovirulent types)	Intestinal tracts of humans and animals	Mild to severe bloody diarrhea, vomiting, cramping, dehydration, and shock. Can result in more serious symptoms. Some illnesses may last up to 8 days.	Raw or rare meats and poultry, raw milk and milk products, unprocessed cheese, salads

13

neurotoxin. Its notable characteristics are its heat-resistant spores and their widespread distribution. Some strains of *C. botulinum* are psychotrophic. The spores survive most thermal processes except those specifically designed to eliminate them (e.g., 12D thermal processing of low-acid canned foods). If such a process is not used, one must assume that spores are present in the food. If the food is to be packaged in an anaerobic or reduced oxygen atmosphere, measures to inhibit the growth and toxin production by the organism are necessary. *C. botulinum* growth can be controlled by one or a combination of the following conditions: pH < 4.6; $a_w \leq 0.94$; 5–10% salt concentration; nitrite and salt combinations (e.g., cured meats); other preservatives; temperature control (freezing/refrigeration), and biocontrol (e.g., inoculation of product with lactic acid bacteria). Sole reliance on refrigeration to ensure safety is risky. Botulinum toxin produced is one of the most potent substances known but is relatively heat labile (destroyed by boiling for 10 minutes). Reliance on final cooking by the consumer to destroy the toxin is extremely risky.

Listeria monocytogenes. *Listeria monocytogenes* is a hazardous foodborne microorganism of relatively recent concern. It is ubiquitious in nature and is commonly found in food processing environments. It causes listerosis, a severe and often fatal illness, to which certain populations (e.g., pregnant mothers, newborns, immunocomprised individuals, transplant recipients) may be susceptible. Fatality rates with the more severe forms of listeriosis can be as high as 70% for those untreated, but generally are between 25 and 35%. The organism is psychotrophic and can grow at refrigeration temperatures. Its widespread distribution and its ability to multiply at refrigeration temperatures and cause severe illness make it a hazard of particular concern to the food industry and regulatory agencies. HACCP programs should attempt to destroy, eliminate, or reduce this hazard and prevent the opportunity for subsequent recontamination.

Salmonella. *Salmonella* species can be found on most raw foods of animal origin. Salmonellosis is one of the most frequently reported foodborne diseases. Symptoms of salmonellosis are most severe in susceptible populations (the elderly, infants, and the infirm). Although about 40,000 cases are reported each year, it is estimated that 2–4 million cases occur annually. *Salmonella* species are destroyed by normal pasteurization processes and are most commonly spread through contamination of processed materials with raw products or with the juices of raw products via hands, utensils, or food-contact surfaces. HACCP plans for processed foods should include controls to destroy and eliminate this organism and to prevent recontamination.

Staphylococcus aureus. *Staphylococcus aureus* may produce a very heat-stable enterotoxin when permitted to grow to an elevated level ($>10^5$ organisms/g). The foodborne intoxication is caused by ingesting enterotoxins produced in food by some strains of *S. aureus,* usually because the food has not been kept hot enough (>60°C; 140°F) or cold enough (≤7.2°C; 45°F or below). The organism is commonly isolated from hands and nasal passages of humans. Thus, foods which are handled or require preparation are at risk. The organism can grow at an a_w of 0.86 and in high salt concentrations. Proper preprocessing handling of raw materials is essential. If conditions allow the organism to grow and produce enterotoxins, subsequent thermal processing will destroy the vegetative organisms while the head-stable toxin persists. There is evidence that the enterotoxins may not be completely inactivated at retort temperatures (121°C or 250°F). HACCP plans should provide for proper handling of raw materials, steps to destroy, eliminate, or reduce the hazard and controls to prevent recontamination. If organisms can reasonably be expected in the final product, conditions to inhibit growth and toxin production should be controlled.

Clostridium perfringens. *Clostridium perfringens* is another anaerobic, spore-forming, rod-shaped bacterium. Perfringens food poisoning is caused by consuming foods that contain large numbers of those *C. perfringens* strains that are capable of producing the food poisoning toxin, which is usually formed in the digestive tract and is associated with sporulation. Limited evidence exists that preformed toxin can be found in food. Perfringens food poisoning is frequently associated with food service operations; temperature abuse of prepared foods, such as large poultry or cooked cuts of meat and gravies and sauces prepared in large containers, can provide anaerobic conditions. Because spores are heat resistant, small numbers of organisms may be present after cooking (or large numbers after improper cooking). Subsequent temperature abuse [not keeping cooked foods above 60°C (140°F) or not providing rapid, even cooling to refrigeration temperatures] may permit the organisms to multiply to food poisoning levels. HACCP plans should control proper cooking conditions and subsequent handling temperatures to inhibit growth of this organism.

Information on the specific limiting growth parameters, heat resistance, growth inhibitors or particular resistance to chemical disinfectants for these and other foodborne bacterial pathogens is available in reference textbooks (Cliver 1990; Doyle 1989; Shapton and Shapton 1991).

Viral hazards

Viruses are very small particles that cannot be seen with a light microscope. They are obligate intracellular parasites that are unable to reproduce outside the host cell. Thus, they are inert in foods and do not multiply in them (Cliver 1988). However, viruses may be transmitted to foods via the fecal-oral route, either directly or indirectly. Some viruses may be inactivated in foods by thorough cooking and some by drying. However, contamination of foods with viruses should be avoided. Direct contamination can occur when an infected food handler contaminates food. Indirect contamination can occur when foods such as bivalve mollusks become contaminated in waters infected by untreated sewage. The viruses most commonly recognized as foodborne disease agents are summarized below.

Hepatitis A virus. Hepatitis A virus (HAV) is classified with the enterovirus group of the *Picornaviridae* family. The terms hepatitis A or type A viral hepatitis have replaced all previous names for the illness. Hepatitis A is usually a mild illness characterized by sudden onset of fever, malaise, nausea, anorexia, and abdominal discomfort, followed in several days by jaundice. Occasionally, the symptoms are severe and convalescence can take several months. The incubation period for hepatitis A varies from 10 to 50 days (mean 30 days). The period of virus shedding or communicability extends from early in the incubation period to about a week after the development of jaundice. The greatest danger of spreading the disease to others occurs 10–14 days before the first presentation of symptoms. The infectious dose is unknown but presumably is 10–100 virus particles.

HAV is excreted in feces of infected individuals and contaminates water or foods via the fecal-oral route. Virtually any food that is handled by an infected worker and not further cooked can serve as a vehicle for transmission. Shellfish (bivalue mollusks), salads, cold cuts and sandwiches, fruits and fruit juices, milk and milk products, vegetables, and iced drinks are commonly implicated in outbreaks. Shellfish (bivalue mollusks), and salads are the most frequent food sources. Virus transmission through foods can best be avoided by preventing fecal contamination and thoroughly cooking foods before consumption.

The Norwalk virus family. Norwalk virus is the prototype of a family of unclassified small round structured viruses (SRSVs) which may be related to the caliciviruses. Common names of the illness caused by the Norwalk and Norwalk-like viruses are viral gastroenteritis and acute nonbacterial gastroenteritis. The disease is self-limiting, mild, and characterized by nausea, vomiting, diarrhea, and abdominal pain. Headache and low-grade fever may occur. The infectious dose is unknown but presumed to be low. Norwalk gastroenteritis is transmitted by the fecal-oral route via contaminated water and foods. Secondary person-to-person transmission has also been documented. Water is the most common source of outbreaks and may include water from municipal supplies, well, recreational lakes, swimming pools, and water stored aboard cruise ships. Salad ingredients and shellfish are the foods most often implicated in Norwalk outbreaks. Ingestion of raw or insufficiently steamed clams and oysters poses a high risk for infection with Norwalk virus. A variety of foods other than shellfish are contaminated by ill food handlers and include salads, fruits, eggs, clams, and bakery items.

Rotavirus. Rotaviruses are classified with the *Reoviridae* family. Rotaviruses cause acute gastroenteritis. Infantile diarrhea, winter diarrhea, acute nonbacterial infectious gastroenteritis, and acute viral gastroenteritis are names applied to the infection caused by the most common and widespread group A rotavirus. Rotavirus gastroenteritis is a self-limiting, mild-to-severe disease characterized by vomiting, watery diarrhea, and low-grade fever. The infective dose is presumed to be 10–100 infectious viral particles. Rotaviruses are transmitted by the fecal-oral route. Infected food handlers may contaminate foods that require handling and no further cooking, such as salads, fruits, and hors d'oeuvres. The virus has not been isolated from any food associated with an outbreak, and no satisfactory method is available for routine analysis of food. Control measures to prevent rotavirus transmission in foods are similar to those used for other viral agents.

Other viruses associated with gastroenteritis. Although the rotavirus and the Norwalk family of viruses are the leading causes of viral gastroenteritis, a number of other viruses have been implicated in outbreaks, including astroviruses, caliciviruses, enteric adenoviruses, and parvovirus. Astroviruses, caliciviruses, and the Norwalk family of viruses possess well-defined surface structures and are sometimes identified as "small round structured viruses" or SRSVs. Viruses with a smooth edge and no discernible surface structure are designated "featureless viruses" or "small round viruses" (SRVs). These agents resemble enterovirus or parvovirus, and may be related to them.

Common names of the illness caused by these viruses are acute nonbacterial infectious gastroenteritis and viral gastroenteritis. Viral gastroenteritis is usually a mild illness characterized by nausea, vomiting, diarrhea, malaise, abdominal pain, headache, and fever. The clinical features are milder but otherwise indistinguishable from rotavirus gastroenteritis. Co-infections with other enteric agents may result in more severe illness lasting a longer period of time. The infectious dose is unknown but is presumed to be low. Viral gastroenteritis is transmitted by the fecal-oral route via person-to-person contact or ingestion of contaminated foods and water. Infected food handlers may contaminate foods that are not further cooked before consumption. Enteric adenovirus may also be transmitted by the respiratory route. Shellfish have been implicated in illness caused by a parvo-like virus.

Parasitic protozoa and worm hazards

Parasites are organisms that derive their sustenance on or within their host. A variety of parasitic animals are of concern to the food microbiologist. They include protozoa, nematodes (roundworms), cestodes (tapeworms), and trematodes (flukes). Table 3-3 lists the parasitic protozoa and worms that are relevant

TABLE 3-3   **Parasites of Major Concern in the United States[a]**

I. Protozoa
   Giardia lamblia
   Entamoeba histolytica
   Cryptosporidium parvum
   Toxoplasma gondii
   Naegleria spp.
   Acanthamoeba spp.

II. Nematodes (roundworms)
   Ascaris lumbricoides
   Trichuris trichiura
   Trichinella spiralis
   Enterobius vermicularis
   Anisakis spp.
   Pseudoterranova spp.

III. Cestodes (tapeworms)
   Taenia saginata
   Taenia solium
   Diphyllobothrium latum

IV. Trematodes (flukes)
   Fasciola hepatica
   Fasciola gigantica

[a](Jackson 1990).

to the food industry in the United States (Jackson 1990). Some foodborne parasites may be transmitted through food and water contaminated by fecal material that contains parasites shed by infected hosts. Other parasites spend a portion of their life cycle in food animals and are thus ingested along with the food. Methods for preventing transmission of parasites to foods via the fecal contamination route include good personal hygiene practices by food handlers, proper disposal of human feces, eliminating the use of insufficiently treated sewage to fertilize crops, and proper sewage treatment. Thorough cooking of foods will eliminate all foodborne parasites. Freezing, and in specific instances brining, may be used to destroy various parasites in foods.

The following overview discusses selected parasitic protozoa and worms. Additional information may be found in these references: Cheng 1986; Cliver 1990; Healy et al. 1984; Jackson 1990.

Giardia lamblia. *Giardia lamblia* (*intestinalis*) is a single-celled protozoa that causes giardiasis in humans. *G. lamblia* exists in two separate stages: the active feeding (trophozoite) stage and the infective environmental (cyst) stage in which the organism survives outside the host. Human giardiasis may involve diarrhea within a week after the cyst is ingested. Other symptoms include abdominal cramps, fatigue, nausea, flatulence, and weight loss. The illness may last for one to two weeks, but chronic infections may last months to years. Colonization and pathogenesis generally occur in the lumen of the small intestine, but the disease mechanism is unknown. *G. lamblia* is shed in the feces of infected individuals and is transmitted via the fecal-oral route. Giardiasis is most frequently associated with the consumption of contaminated water. Outbreaks have been traced to food contamination by infected food handlers, and the possibility of infection from contaminated vegetables that are eaten raw cannot be excluded. Cool moist conditions favor the survival of the organism. Food contamination by infected food workers can be prevented by proper personal hygiene. Thorough cooking of foods destroys *G. lamblia*.

Entamoeba histolytica. *Entamoeba histolytica* is a single-celled protozoa that predominantly infects humans and other primates. Like *G. lamblia*, *E. histolytica* can exist as two separate stages: a trophozoite or a cyst. Cysts survive outside in water, in soils, and on foods, especially under moist conditions. When swallowed, they cause infections by excysting (to the trophozoite stage) in the digestive tract. Infections can be asymptomatic or accompanied by a mild gastrointestinal distress or dysentery (with blood and mucus). *E. histolytica* may penetrate the intestinal wall, and if it enters the blood, may gain access to other organs. Large numbers of cysts can be shed in the feces of infected individuals. *E. histolytica* is transmitted via the fecal-oral route. Infection can result from the fecal contamination of drinking water and foods, and by direct contact with dirty hands or objects. Preventive measures are similar to those described for *G. lamblia*.

Ascaris lumbricoides. Humans worldwide are infected with *Ascaris lumbricoides*. The eggs of this roundworm (nematode) are "sticky" and may be carried to the mouths by hands, other body parts, fomites (inanimate objects), or foods. Ascariasis, the scientific name for this infection, is also commonly known as the "large roundworm" infection. Ingested eggs hatch in the intestine, and larvae begin to migrate, reaching the lungs through the blood and lymph systems. In the lungs, the larvae break out of the pulmonary capillaries into the air sacs, ascend into the throat, and descend again to the small intestine where they grow to sexual maturity. On occasion, larvae will crawl up into the throat and try to exit through the mouth or nose. Vague digestive tract discomfort sometimes accompanies the intestinal infection, but intestinal blockage may occur in small children who have more than a few worms because of the large size of the worms. Large numbers of eggs may be voided in feces. *A. lumbricoides* eggs are extremely resistant to sewage treatment and may survive in soil for years. The eggs are found in insufficiently treated sewage fertilizer and in soils where they embryonate (i.e., larvae develop in fertilized eggs). The eggs may contaminate crops grown in soil or fertilized with sewage that has received nonlethal treatment. Humans are infected when such produce is consumed raw. Infected foodhandlers may contaminate a wide variety of foods. Careful disposal of human feces and avoiding fertilization of crops with insufficiently treated sewage are key preventive measures. Eggs are slightly susceptible to drying and begin losing their infectivity at temperatures above 38°C (Cliver 1990).

Diphyllobothrium latum. *Diphyllobothrium latum* and other members of the genus are broad fish tapeworms (cestodes). *D. latum* is a broad, long tapeworm, often growing to lengths between 1 and 2 meters (3–7 feet) and potentially capable of attaining 10 meters (32 feet). The disease caused by broad fish tapeworm infections is called diphyllobothriasis. Infections are acquired by consumption of raw, underprocessed or lightly cooked fish. Freshwater fish (e.g., pike, burbot, and perch) and those that migrate between ocean and fresh waters (salmonid fishes) may be infected. The larvae that infect people (plerocercoid) are frequently encountered in the viscera of freshwater and marine fishes. The ingested plerocercoid develops into a mature adult tapeworm and attaches itself to the intestinal wall. Diphyllobothriasis is characterized by abdominal distention, flatulence, intermittent abdominal cramping, and diarrhea with onset about 10 days after consumption of raw or insufficiently cooked fish. The tapeworm has a strong affinity for vitamin B_{12} and may cause a deficiency in the host. In regions where raw or lightly cooked fish are eaten, the frequency of the disease tends to be high. Preventive measures call for adequate cooking of fish foods. Other methods proposed for destroying the larvae in infected fish include freezing or brining at high salt concentrations.

CHEMICAL HAZARDS

Webster defines a chemical as any substance used in or obtained by a chemical process or processes. All food products are made up of chemicals, and all chemicals can be toxic at some dosage level. However, a number of chemicals are not allowed in food and others have established allowable limits. A summary of most of the chemical hazards in foods has been compiled (Bryan 1984). The two types of chemical hazards in foods are naturally occurring and added chemicals (Table 3-4). Both may potentially cause chemical intoxications if excessive levels are present in a food. For additional information, see Chemical Intoxi-

TABLE 3-4   Types of Chemical Hazards

I. Naturally occuring chemicals
    Mycotoxins (e.g., aflatoxin)
    Scombrotoxin (histamine)
    Ciguatoxin
    Mushroom toxins
    Shellfish toxins
        Paralytic shellfish poisoning (PSP)
        Diarrheic shellfish poisoning (DSP)
        Neurotoxic shellfish poisoning (NSP)
        Amnesic shellfish poisoning (ASP)
    Pyrrolizidine alkaloids
    Phytohemagglutinin   .
    Polychlorinated biphenyls (PCBs)

II. Added chemicals
    Agricultural chemical
        Pesticides, fungicides, fertilizers, insecticides,
            antibiotics and growth hormones
    Prohibited substances (21 CFR, Section 189)
        Direct
        Indirect
    Toxin elements and compounds
        Lead, zinc, arsenic, mercury and cyanide
    Food additives
        Direct—allowable limits under GMPs
            Preservatives (nitrite and sulfiting agents)
            Flavor enhancers (monosodium glutamate)
            Nutritional additives (niacin)
            Color additives
        Secondary direct and indirect
            Plant chemicals (e.g., lubricants, cleaners,
                sanitizers, cleaning compounds, coating
                and paint)
        Chemicals intentionally added (sabotage)

cations and Naturally Occurring Toxicants in Foods in *Foodborne Diseases* (Cliver 1990). Many HACCP programs have been criticized for their relative neglect of chemical and physical hazards.

Naturally occurring chemicals

If formal limits have been established for naturally occurring toxicants, the limit will be found in the Code of Federal Regulations, Title 21. If informal limits have been established (e.g., aflatoxins, paralytic shellfish toxin, and scombrotoxin), the maximum allowable limit will be found in the Food and Drug Administration's (FDA) Compliance Policy Guidelines, available from FDA, Center for Food Safety and Applied Nutrition HFF-300, Washington, DC 20204. The naturally occurring toxicants include a variety of chemicals of plant, animal, or microbial origin. Although many naturally occurring toxicants are biological in origin, they have traditionally been categorized as chemical hazards. However, for individual HACCP programs, their inclusion in the biological hazard category would be equally appropriate. The following overview discusses several naturally occurring toxicants.

Mycotoxins. A number of fungi produce compounds (mycotoxins) toxic to man (Stoloff 1984). Mycotoxins are secondary metabolites of certain fungi. Among some of the better known and studied groups of mycotoxins are the aflatoxins, which include a group of structurally related toxic compounds produced by certain strains of the fungi *Aspergillus flavus* and *A. parasiticus*. Under favorable conditions of temperature and humidity, these fungi grow and produce aflatoxins on certain foods, grains, nuts, and feeds. The most pronounced contamination has been encountered in tree nuts, peanuts, and other oilseeds including corn and cottonseed. The major aflatoxins of concern are designated B_1, B_2, G_1, and G_2, which are usually found together in some foods and feeds in varying proportions. However, aflatoxin B_1 usually predominates and is the most toxic. In the United States, aflatoxins have been identified in corn and corn products, peanuts and peanut products, cottonseed, milk, animal feeds, and tree nuts such as Brazil nuts, pecans, pistachio nuts, and walnuts. Other grains and nuts are susceptible, but are less prone to contamination.

Scombrotoxin (Histamine). Scombroid poisoning or histamine poisoning occurs when foods that contain high levels of histamine (or possibly other vasoactive amines and compounds) are ingested. Histamine is produced by the microbial degradation of histidine, a free amino acid found in abundance in dark-fleshed fish, including members of the *Scombridae* family from temperate and tropical regions. Fish that have been temperature abused are the most commonly implicated foods. Other foods such as Swiss cheese have been reported to cause illness as well. Fish most often implicated are mahi mahi, tuna, mackerel, bluefish, and amberjack.

Ciguatera. Ciguatera is a form of human poisoning caused by the consumption of subtropical and tropical marine finfish which have accumulated naturally occurring toxins through their diet. The toxins originate from several dinoflagellate (algae) species common to ciguatera endemic regions and accumulate through the food chain. Manifestations of ciguatera in humans usually involves a combination of gastrointestinal, neurological, and cardiovascular disorders.

Marine finfish most commonly implicated in ciguatera fish poisoning are predators and include the groupers, barracudas, snappers, jacks, mackerel, and triggerfish. Other species of warm-water fishes have been reported to harbor ciguatera toxins. The presence of toxic fish is sporadic; not all fish from a given locality or species will be toxic.

Mushroom toxins. Mushroom poisoning is caused by the consumption of raw or cooked fruiting bodies of certain higher fungi. Unlike the previously mentioned aflatoxins, which are secondary metabolites produced when a contaminating mold grows on a food product, the mushroom itself is the toxic food product. Many species of mushrooms are toxic and there is no general rule to distinguish between edible and toxic species. Mushroom poisonings are usually caused by ingestion of toxic wild mushrooms that have been confused with edible species. Most mushrooms that cause human poisoning cannot be rendered nontoxic by cooking, canning, or freezing.

Shellfish toxins. Shellfish poisoning is caused by a group of toxins elaborated by planktonic algae (dinoflagellates, in most cases) upon which the shellfish feed. Under the appropriate conditions toxic dinoflagellate populations may increase to high levels and persist for several weeks. The shellfish may accumulate and metabolize these toxins during their filter feeding. There are four types of shellfish poisonings: paralytic shellfish poisoning (PSP); diarrheic shellfish poisoning (DSP); neurotoxic shellfish poisoning (NSP); and amnesic shellfish poisoning (ASP). Ingestion of contaminated shellfish results in a wide variety of symptoms, depending upon the toxin(s) present, their concentrations in the shellfish, and the amount of contaminated shellfish consumed (Hall 1991). All shellfish (filter-feeding mollusks) could potentially become toxic. However, PSP is generally associated with mussels, clams, cockles, and scallops; NSP with shellfish harvested along the Florida coast and the Gulf of Mexico; DSP with mussels, oysters, and scallops; and ASP with mussels. Control methods include effective monitoring of shellfish lots or growing areas and, in some instances, depuration.

The food processor may control some of these naturally occurring chemical hazards by learning in which foods (i.e., sensitive ingredient) they are most likely to occur. Proper raw material specification, vendor certification, and guarantees along with inspection and spot checks will help to prevent introduction of natural chemical hazards into plant facilities. Likewise, proper handling and storage of sensitive ingredients will prevent conditions conducive to the production of other natural toxicants (e.g., proper storage of grains and feeds to prevent aflatoxin production and avoidance of temperature abuse of fish susceptible to scombroid poisoning).

Added Chemicals

The second group of chemicals which may be potential hazards are those that are added to foods at some point between growing, harvesting, processing, storage, and distribution (Table 3-4). These chemicals are generally not considered hazardous if proper conditions of use are followed. Only when these chemicals are misapplied or when their permitted levels are exceeded is there a potential hazard. Cliver (1990) has reviewed the added chemical hazards. The first group of added chemicals includes agricultural chemicals, such as pesticides, herbicides, fungicides, fertilizers, antibiotics, and growth hormones. Pesticides and herbicides are regulated by the Environmental Protection Agency (EPA), which specifically states the permitted uses of each chemical and the maximum allowable residue levels.

Prohibited substances (Table 3-4) are listed in Title 21, Part 189, of the Code of Federal Regulations. Their direct or indirect use in food is prohibited because they present a potential risk to the public health or have not been shown by adequate scientific data to be safe for use in human food.

Toxic elements (e.g., lead, mercury, arsenic) and other toxic compounds (e.g., some chemicals used in the food processing plant) are either not allowed in food at all or have established maximum tolerances. In some cases these chemicals are present naturally and have not been added to the food. Additional reference information on many of these toxic elements can be found in *Handbook on the Toxicology of Metals* (Friberg, Nordberg and Vouk 1979). Other added chemicals in the food additive group, including direct, secondary direct, and indirect food and color additives, are permitted to be used in actual food processing to preserve the food (e.g., preservatives), enhance flavor, impart color, or nutritionally fortify (e.g., vitamins and minerals). Secondary direct and indirect chemicals used in food processing plants include chemicals such as lubricants, cleaners, sanitizers, paint, and coatings, which may become incorporated into food via migration from packaging materials, or microorganisms and enzyme preparations used in food processing. Allowable limits for all of these food additives have been set in accordance with Good Manufacturing Practices (GMPs). At established limits these chemicals are not hazardous and a large safety factor is incorporated into the regulatory limits; however, if tolerances are exceeded, potential health risks to consumers may occur.

Means of control of chemical hazards are listed in Table 3-5. Foods which contain levels of agricultural chemicals exceeding permitted tolerances should not be accepted. Proper raw material specifications, vendor certification, and guarantees along with inspection and spot checks will help to prevent the introduction or receipt of added chemical hazards in food material. Other chemicals should be checked for intended uses, purity, formulation, and proper labeling. Quantities of chemicals to be added to foods or used in food processing areas

TABLE 3-5   Control of Chemical Hazards

I. Control before receipt
    Raw material specifications
    Vendor certification/guarantees
    Spot checks—vertification

II. Control before use
    Review purpose for use of chemical
    Ensure proper purity, formulation and labeling
    Control quantities to be added

III. Control storage and handling conditions
    Prevent conditions conducive to production of naturally occurring toxicants

IV. Inventory all chemicals in facility
    Review uses
    Records of use

must be controlled and recorded. Premeasured quantities of food additives (e.g., preservatives, nitrites, nutritional enhancers, color additives) can be prepared ahead of time and properly labeled or color-coded to avoid confusion in the absence of supervision. Methods to prevent intentionally added chemical hazards are similar to those prescribed for preventing intentionally added physical hazards.

PHYSICAL HAZARDS

Physical hazards are often described as extraneous matter or foreign objects and include any physical matter not normally found in food which may cause illness (including psychological trauma) or injury to an individual (Corlett 1991). The FDA maintains a passive surveillance system known as the Complaint Reporting System for the reporting of consumer complaints related to food items. A total of 10,923 complaints regarding food items consumed during the period October 1, 1988, through September 30, 1989, were reported to the FDA Complaint Reporting System (Hyman, Klontz and Tollefson 1991). The largest single category (2,726 complaints) involved the presence of foreign objects in food and accounted for 25% of all complaints. Of all reported foreign object complaints, 387 (14%) resulted in illness or injury. The most common foreign object in those reports was glass. Table 3-6 lists the most frequently implicated food types involved in foreign object complaints. One reason physical hazards are the most often reported complaint is that foreign objects provide tangible evidence of a product deficiency. Regulatory action may be initiated when agencies find adulterated foods or foods that are manufactured, packed or held under conditions whereby they may have become contaminated or rendered injurious to health. Thus, although the discovery of filth in a product may not itself present an unacceptable heath risk, the conditions of manufacture, packaging, or storage that permitted its entry present an unacceptable health risk. Food processors must therefore be aware of product adulteration by physical substances and address their control in a HACCP program.

The main physical hazards of concern, their sources, and injury potential are listed in Table 3-7. This list is by no means all inclusive; almost anything imaginable can ultimately be introduced into food and present a physical hazard. Other items not mentioned in Table 3-7 include hair, dirt, paint and paint chips, rust, grease, dust, and paper. The sources of physical hazards include raw materials, water, facility grounds, equipment, building materials, and employee personal effects. Physical hazards may be added inadvertently during distribution and storage, or intentionally introduced (sabotage).

Methods involved in controlling physical hazards include raw material specifications and inspections along with vendor certification and guarantees. Various preventive measures are available to find and remove certain physical hazards. Metal detectors can be used to locate ferrous and nonferrous metals in foods; various foreign objects, especially bone fragments can be found through X-ray technology. Effective pest control and foreign object removal from plant environments are also essential. Preventive maintenance and sanitation programs for plants and equipment are necessary. Proper shipping, receiving, distribution and

TABLE 3-6   Eight Most Common Food
Categories Implicated in Reported
Foreign Object Complaints[a]

Food Category	Number of Complaints	Percent[b]
Bakery	277	10.2
Soft drinks	228	8.4
Vegetables	226	8.3
Infant foods	187	6.9
Fruits	183	6.7
Cereal	180	6.6
Fishery	145	5.3
Chocolate and cocoa products	132	4.8

[a]Adapted from Hyman et al. (1991). Does not include meat and poultry categories or suspected or confirmed tampering complaints.
[b]Percent of total (2,726) reported foreign object complaints received by the FDA Complaint Reporting System from 10/1/88 through 9/30/89.

TABLE 3-7   **Main Materials of Concern as Physical Hazards and Common Sources[a]**

Material	Injury Potential	Sources
Glass	Cuts, bleeding; may require surgery to find or remove	Bottles, jars, light fixtures, utensils, gauge covers
Wood	Cuts, infection, choking; may require surgery to remove	Fields, pallets, boxes, buildings
Stones	Choking, broken teeth	Fields, buildings
Metal	Cuts, infection; may require surgery to remove	Machinery, fields, wire, employees
Insects and other filth	Illness, trauma, choking	Fields, plant post-process entry
Insulation	Choking; long-term if asbestos	Building materials
Bone	Choking, trauma	Fields, improper plant processing
Plastic	Choking, cuts, infection; may require surgery to remove	Fields, plant packaging materials, pallets, employees
Personal effects	Choking, cuts, broken teeth; may require surgery to remove	Employees

[a]Adapted from Corlett (1991).

storage procedures as well as packaging material handling practices (particularly those involving glass) must be evaluated for their potential to introduce hazards. Packaging should be tamper-proof and at least tamper-evident. Finally, employee education and practices must involve knowledge and prevention of physical hazard introduction.

ACKNOWLEDGMENTS

The author thanks the many FDA scientists who contributed to the internal FDA document *Foodborne Pathogenic Microorganisms and Natural Toxins,* from which much of the information in this chapter was obtained.

References

Archer, D.L. and Kvenberg, J.E. 1985. Incidence and cost of foodborne diarrheal disease in the United States. J. Food Prot. 48: 887–894.

Bryan, F.L. 1979. Epidemiology of foodborne diseases. In *Food-Borne Infections and Intoxications.* (Ed.) H. Riemann and F.L. Bryan, p. 4–69. Academic Press, New York.

Bryan, F.L. 1984. Appendix: Diseases transmitted by foods. In *Adverse Reactions to Foods.* (Ed.) J.A. Anderson and D.D. Sogn, p. 1–101. U.S. Dept. of Health and Human Services, NIH Publication No. 84-2442, Washington, DC.

Cheng, T.C. 1986. *General Parasitology.* Academic Press, Orlando.

Cliver, D.O. 1988. Virus transmission via foods; A scientific status summary by the Institute of Food Technologists' Expert Panel on Food Safety and Nutrition. Food Technol. 42(10): 241–248.

Cliver, D.O. 1990. *Foodborne Diseases*. Academic Press, San Diego, CA.

Corlett, D.A. 1991. A practical approach to HACPP. Food Safety Management Seminar. ESCAgenetics Corp., 830 Bransten Rd., San Carlos, CA.

Doyle, M.P. (Ed.). 1989. *Foodborne Bacterial Pathogens*. Marcel Dekker, Inc., New York.

Friberg, L., Nordberg, G. F., and Vouk, V. B. 1979. *Handbook on the Toxicology of Metals*. Elsevier/North-Holland, Amsterdam.

Hall, S. 1991. Natural Toxins. In *Microbiology of Marine Food Products*. (Ed.) D. Ward and C. Hackney, p. 301–330. Van Nostrand Reinhold, New York.

Healy, G.R., Jackson, G.J., Lichtenfels, J.R., Hoffman, G.L., and Cheng, T.C. 1984. Foodborne parasites. In *Compendium of Methods for the Microbiological Examination of Foods*. (Ed.) M.L. Speck, p. 542–556. American Public Health Assoc., Washington, DC.

Hyman, F.N., Klontz, K.C., and Tollefson, L. 1991. The role of foreign objects in the cause of foodborne injuries: Surveillance by the Food & Drug Administration. (Submitted for publication.)

International Commission on Microbiological Specifications for Foods. (ICMSF). 1986. *Microorganisms in Foods*. Vol. 2. Univ. of Toronto Press, Toronto.

Jackson, G.L. 1990. Parasitic protozoa and worms relevant to the U.S. *Food Technol.* 44(5): 106–1112.

Riemann, H. and Bryan, F.L. 1979. *Foodborne Infections and Intoxications*. Academic Press, New York.

Shapton, D.A. and Shapton, N.F. *Principles and Practices for the Safe Processing of Foods*. Butterworth-Heinemann Ltd., Oxford.

Stoloff, L. 1984. Toxigenic fungi. In *Compendium of Methods for the Microbiological Examination of Foods*. (Ed.) M.L. Speck, p. 557–572. American Public Health Assoc., Washington, DC.

Todd, E.C.D. 1989. Preliminary estimates of costs of foodborne disease in the United States. J. Food Prot. 52: 595–601.

Ward, D. and Hackney, C. 1991. *Microbiology of Marine Food Products*. Van Nostrand Reinhold, New York.

4

Hazard Analysis and Assignment of Risk Categories

Donald A. Corlett, Jr.
Merle D. Pierson

Principle 1. *Assess hazards associated with growing, harvesting, raw materials and ingredients, processing, manufacturing, distribution, marketing, preparation and consumption of the food.*

OVERVIEW

Hazard analysis consists of a systematic evaluation of a specific food and its raw materials or ingredients to determine the risk from biological (primarily infectious or toxin-producing food-borne illness microorganisms), chemical and physical hazards. The hazard analysis is a two-step procedure: hazard analysis and assignment of risk categories.

The first step is to rank the food and its raw materials or ingredients according to six hazard characteristics (A-F). A food is scored by using a plus (+) if the food has the characteristic, and a zero (0), if it does not exhibit the characteristic. The six characteristic ranking system is applied for microbiological, chemical and physical hazard ranking, although the characteristics are somewhat different for microbiological and chemical/physical hazards, as described later.

The second step is to assign *risk categories* (VI) to the food and its raw materials and ingredients based on the results of ranking by hazard characteristics. Possible combinations of hazard characteristic ranking and hazard categories are presented in Table 4-1. Potentially highest risk is denoted by the highest number in the hazard category (i.e., VI). In addition, note that whenever there is a plus

This entire section is adapted from the training course, "A Practical Application of HACCP," copyrighted, 1990, by ESCAgenetics Corporation and licensed to D.A. Corlett. Permission is granted for the Institute of Food Technologists to reproduce this material for the 1991 IFT Short Course: Hazard Analysis and Critical Control Points.

TABLE 4-1   **Possible Combinations of Hazard Characteristic Ranking and Hazard Categories for Food Products and Food Raw Materials and Ingredients**

Food Ingredient or Product[a]	Hazard Characteristics (A, B, C, D, E, F)	Risk Category
T	A + (Special category)[b]	VI
U	Five + 's (B through F)	V
V	Four + 's (B through F)	IV
W	Three + 's (B through F)	III
X	Two + 's (B through F)	II
Y	One + (B through F)	I
Z	No + 's	0

[a]The letters merely indicate different types of foods having different hazard characteristics and risk categories. Normally the name of a food, raw material or ingredient would appear under this heading.
[b]Hazard characteristic A automatically is risk category VI, but any combination of B through F may also be present.

(+) for hazard characteristic A (a special class that applies to food designated for high-risk populations), the resulting hazard category is always VI, even though other hazard characteristics (B-F) may or may not be a plus (+).

Several preliminary steps are needed before conducting the hazard analysis. These include developing a working description of the product, listing the raw materials and ingredients required for producing the product, and preparation of a diagram of the complete food production sequence. The listing of raw materials and ingredients is the starting point for the hazard analysis. If the specific mode of preservation for an ingredient is not known (raw, frozen, canned, etc.), the ingredient may be assessed for each type of preservation technique that may be utilized in preserving the ingredient.

The following parts of this section cover the details of the hazard analysis for microbiological, chemical and physical hazards, and illustrate the application of hazard characteristics and assignment of hazard categories for various foods.

MICROBIOLOGICAL HAZARD CHARACTERISTIC RANKING

Microbiological hazard analysis and the ranking of food by hazard characteristics is explained in detail in Section 4.1, pages 3–5 of the pamphlet, HACCP Principles for Food Production (NACMCF 1989). I have made several minor changes in Hazard F, to differentiate ranking for consumer products, and raw materials and ingredients as received by the processor before any manufacturing steps.

TABLE 4-2   Microbiological Risk Characteristics[a]

Hazard A:	A special class that applies to nonsterile products designated and intended for consumption by at-risk populations (e.g., infants, the aged, the infirm, or immunocompromised individuals).
Hazard B:	The product contains "sensitive ingredients" in terms of microbiological hazards.
Hazard C:	The process does not contain a controlled processing step that effectively destroys harmful microorganisms.
Hazard D:	The product is subject to recontamination after processing before packaging.
Hazard E:	There is substantial potential for abusive handling in distribution or in consumer handling that could render the product harmful when consumed.
Hazard F:	There is no terminal heat process after packaging or when cooked in the home. *(Applies to food product, as used by the consumer.)* There is no terminal heat process or any other kill-step applied after packaging by the vendor, or other kill-step applied before entering food manufacturing facility. *(Applies to raw materials and ingredients coming into a food manufacturing facility.)*

[a]After NACMCF HACCP system (USDA-FSIS, 1990); and by permission of D. Corlett (Copyright D. Corlett by license from ESCAgenetics Corporation, course manual, *A Practical Application of HACCP*, 1990).

The microbiological hazard characteristics are given in Table 4-2. As indicated earlier, rank the product and its raw materials and ingredients according to hazard characteristics A through F, using a plus (+) to indicate that the food product or its raw materials or ingredients exhibit the characteristic, and a zero (0) when they do not.

A brief discussion of "microbiologically sensitive" products, and raw materials and ingredients, is useful for scoring foods for Hazard Characteristic B given on Table 4-2. Give the *product* a plus (+) if it is sensitive or contains micro-biologically sensitive ingredient(s). Give *raw materials* or *ingredients* a plus (+) if they are microbiologically sensitive or contain sensitive foods (e.g., a cheese/starch flavor blend).

A "sensitive ingredient" is defined as "any ingredient historically associated with a known microbiological hazard." The term "ingredient" normally also applies to raw materials. "Sensitive ingredient" was coined for microbiological hazards (infectious agents and their toxins), but it is also now used for ingredients and raw materials that are historically associated with known chemical or physical hazards.

The original list of microbiologically sensitive foods was based on the potential presence of the *Salmonella* species. Now any type of hazardous microorganism may cause a food to be "sensitive," and the list of sensitive foods has grown, particularly with the recognition that *Listeria monocytogenes* is a known threat in many foods. A partial listing of sensitive raw materials and ingredients is provided in Table 4-3 to assist in scoring a food, or its raw materials and

**TABLE 4-3   Microbiologically Sensitive
Raw Materials and Ingredients**

Meat and poultry
Eggs
Milk and dairy products (including cheese)
Fish and shellfish
Nuts and nut ingredients
Spices
Chocolate and cocoa
Mushrooms
Soy flour and related materials
Gelatin
Pasta
Frog legs
Vegetables
Whole grains and flour (secondary contamination)
Yeast
Dairy cultures
Some colors and flavors from natural sources

ingredients, for Hazard Characteristic B. If there is a question as to whether a food is sensitive, it should be considered sensitive until more information is available for purposes of clarifying its status.

Compounded ingredients may be considered sensitive if they are combinations of sensitive and nonsensitive ingredients. For example, a fat coated on milk powder, or compounded cheese flavor coated on starch. It is best to list all components of a blended material to determine if the blend contains a sensitive ingredient and also determine if it has received a controlled processing step that destroys hazardous microorganisms. In some cases, it is important to determine if microbiological toxins may also be present in a "processed" food, if it is to

**TABLE 4-4   Foods Not Normally Considered
Sensitive**

Salt
Sugar
Chemical preservatives
Food grade acidulents and leavening agents
Gums and thickeners (some may be sensitive depending on
    origin, such as tapioca and fermentation-derived gums)
Synthetic colors
Food grade antioxidants
Acidified high salt/acid condiments
Most fats and oils (exception is dairy butter)
Acidic fruits

be used as an ingredient (e.g., heat stable staphylococcus enterotoxin in canned mushrooms).

Many raw materials and ingredients are not considered microbiologically sensitive even though they may occasionally be contaminated with hazardous microorganisms. A partial list is included in Table 4-4.

Tables 4-3 and 4-4 are not necessarily an exhaustive listing of all sensitive and nonsensitive ingredients. When in doubt, it is recommended that assistance be obtained from authoritative sources including universities, regulatory agencies, trade organizations, consultants and consulting laboratories.

CHEMICAL AND PHYSICAL HAZARD RISK ASSESSMENT PROCEDURES

The following protocol for hazard analysis of chemical and physical food hazards complements the existing microbiological hazard analysis scheme given in the

TABLE 4-5   Hazard Characteristics for Ranking foods for Chemical and Physical Hazards[a]

HAZARD A:	A special class that applies to products designated and intended for consumption by high-risk populations (e.g., infants, the aged, the infirm, or immunocompromised individuals). (Examples are foods intended for persons sensitive to sulfites, and for infants where glass is of particular concern.)
HAZARD B:	The product contains "sensitive" ingredients known to be potential sources of toxic chemicals or dangerous physical hazards. (Examples are aflatoxin in field corn, and stones in agricultural products.)
HAZARD C:	The process does not contain a controlled step that effectively prevents, destroys or removes toxic chemical or physical hazards. (Examples include steps for prevention of the formation of toxic or carcinogenic substances during processing; destruction of cyanide-containing compounds by roasting of apricot pits; and removal of toxic processing chemicals such as lye or dangerous foreign objects such as sharp pieces of metal.)
HAZARD D:	The product is subject to recontamination after manufacturing before packaging. (Example is where contamination may occur when a manufactured product is bulk packed, shipped and packaged in another facility.)
HAZARD E:	There is substantial potential for chemical or physical contamination in distribution or in consumer handling that could render the product harmful when consumed. (Examples are contamination of a food from containers or vehicle compartments that previously contained toxic chemicals or foreign objects; selling food in open containers; or where the potential for product tampering is high.)
HAZARD F:	There is no way for the consumer to detect, remove or destroy a toxic chemical or dangerous physical agent. (Examples are presence of toxic mushrooms or paralytic shellfish toxins, or presence of sharp metal objects buried in a food.)

[a]Abstracted from Corlett and Stier (1991).

NACMCF system. Hazard characteristics for chemical and physical agents were developed in 1990 for use in the ESCAgenetics Corporation training course, "A Practical Application of HACCP," and were recently published (Corlett and Stier, 1991). They are designed so that both chemical and physical hazards in food may be assessed by using the same six hazard characteristics.

Generally, hazard analysis for chemical and physical hazards is conducted like the procedure for microbiological hazards provided in the NACMCF guide. Although the six hazard characteristics are somewhat different, the same plus (+) and zero (0) scoring system and hazard category assignment procedures are used.

Table 4-5 provides the hazard characteristics for ranking foods for both chemical and physical hazards. This table also includes examples of chemical and physical agents that could potentially be present in a food relative to each hazard characteristic. The concept of "sensitive" products, raw materials and ingredients is also used in Hazard Characteristic B for chemical and physical hazards.

EXAMPLES OF THE COMBINED HAZARD ANALYSIS FOR CHEESE DIP

The complete hazard analysis consisting of ranking of potential microbiological, chemical and physical hazards and assignment of hazard categories is illustrated in the example of a hypothetical cheese dip product (called Don's Delight, Table 4-6).

TABLE 4-6   Cheese Dip Ingredients (Don's Delight)[a]

	Types of Potential Hazards		
Ingredient	Microbiological	Chemical	Physical
Raw celery	*Salmonella* sp.	Pesticides	Metal
	Shigella sp.		Wood
	Listeria monocytogenes		Rocks
Dried mushrooms	*Salmonella* sp.	Pesticides	Metal
	Shigella sp.		Wood
	Staphylococcus aureus		Rocks
Soft-ripened cheese	*Listeria monocytogenes*	Pesticides	Metal
	Salmonella sp.	Antibiotics	
	Staphylococcus aureus	Hormones	
	EP *Escherichia coli*		
Water	Microbial pathogens	Various	n/u
Salt	Not usually (n/u)	n/u	Metal
Stabilizer	Not usually (n/u)	n/u	Metal

[a]From ECSAgenetics Corporation course "A Practical Application of HACCP".

HACCP PRINCIPLE 1.   HACCP WORKSHEET   FORM 5.0

RISK ASSESSMENT WORK-SHEET FOR MICROBIOLOGICAL FOOD HAZARDS

PRODUCT:___CHEESE DIP_____PAGE_1_OF__1__PAGES DATE:_____
           (DON'S DELIGHT)

FOOD PRODUCT(S)...........AS USED BY THE CONSUMER........................

PRODUCT	MICROBIOLOGICAL HAZARD CHARACTERISTICS ASSOCIATED WITH THE FOOD (+ FOR "YES"; O FOR "NO")						
	A HIGH RISK SPECIAL POPULAT.	B SENSITIVE INGRED-IENTS	C NO KILL-STEP IN PROCESS	D RECONTAM. BETWEEN PROC/PACK	E ABUSIVE HANDLING DIST/CONS	F NO TERM. HEAT PROC BY CONSUM	HAZARD CATEG.
(1) REFRIG.	O	+	+	+	+	+	V.
(2) FROZEN	O	+	O	+	+	+	IV.
(3) CANNED	O	+	O	O	O	+	II.

RAW MATERIALS AND INGREDIENTS...AS RECEIVED, BEFORE ANY MANUFACTURING STEPS
                            BY THE FOOD FACILITY (SUCH AS COOKING)......

RAW MAT. OR INGRE.	A	B	C	D	E	F:NO KILL STEP BEFORE RECEIPT*	HAZARD CATEG.
RAW CELERY	O	+	+	+	+	+	V.
DRIED MUSHROOMS	O	+	+	+	O	+	IV.
SOFT-RIPENED CHEESE	+	+	+	+	+	+	V.
SALT	O	O	O	O	O	O	O.
WATER	O	+	O	+	O	+	III.
STABILIZER	O	O	O	O	O	O	O.

* NO HEAT PROCESS OR ANY OTHER KILL-STEP APPLIED AFTER PACKAGING BY
  SUPPLIER; NO HEAT PROCESS OR OTHER KILL-STEP BEFORE ENTERING FOOD PLANT.

Copyright 1990 by ESCAgenetics Corporation and licensed to D.A. Corlett.
DONSMICR
                                                                4-10

FIGURE 4-1.   Form 5.0. Risk assessment work sheet for microbiological food hazards, HACCP
Principle 1.

```
HACCP PRINCIPLE 1.            HACCP WORKSHEET            FORM 6.0-A

RISK ASSESSMENT WORK-SHEET FOR CHEMICAL OR PHYSICAL FOOD HAZARDS

IS THIS SHEET TO BE USED FOR CHEMICAL OR PHYSICAL HAZARDS?___"CHEMICAL"_____

PRODUCT:___CHEESE DIP (DON'S DELIGHT)_____DATE:_____
```

FOOD ITEM	HAZARD CHARACTERISTICS KNOWN TO BE ASSOCIATED WITH THE FOOD AND IT'S INGREDIENTS (+ FOR "YES"; O FOR "NO")						HAZARD CATEG.
(1) PRODUCT	A HIGH RISK SPECIAL POPULAT.	B INGREDS. CONTAIN HAZARD	C NOT RE- MOVED IN MANUFACT.	D RECONTAM. BETWEEN MAN./PAC.	E CONTAM. BY DIST. OR CONS.	F CONS.CAN- NOT DE- TECT/REM.	
REFRIGERATED	O	+	+	+	O	+	IV.
FROZEN	O	+	+	+	O	+	IV.
CANNED	O	+	+	+	O	+	IV.
(2) RAW MAT'S AND ING'S							
RAW CELERY	O	+	+	+	+	+	V.
DRIED MUSHROOMS	O	+	+	+	O	+	IV.
SOFT-RIPENED CHEESE	+	+	+	+	O	+	IV.
SALT	O	O	O	O	O	O	O.
WATER	O	+	+	O	O	+	III.
STABILIZER	O	O	O	+	O	+	II.

```
NOTES: (1) AS USED BY CONSUMER
       (2) AS ENTERING THE FOOD FACILITY BEFORE PREPARATION OR PROCESSING
```

4-11

FIGURE 4-2.   Form 6.0-A. Risk assessment work sheet for chemical food hazards, HACCP Principle 1.

HACCP PRINCIPLE 1.   HACCP WORKSHEET   FORM 6.0-B

RISK ASSESSMENT WORK-SHEET FOR CHEMICAL OR PHYSICAL FOOD HAZARDS

IS THIS SHEET TO BE USED FOR CHEMICAL OR PHYSICAL HAZARDS?__"PHYSICAL"_____

PRODUCT:___CHEESE DIP (DON'S DELIGHT)_____DATE:_____

FOOD ITEM	HAZARD CHARACTERISTICS KNOWN TO BE ASSOCIATED WITH THE FOOD AND IT'S INGREDIENTS (+ FOR "YES"; O FOR "NO")						HAZARD CATEG.
(1) PRODUCT	A HIGH RISK SPECIAL POPULAT.	B INGREDS. CONTAIN HAZARD	C NOT RE-MOVED IN MANUFACT.	D RECONTAM. BETWEEN MAN./PAC.	E CONTAM. BY DIST. OR CONS.	F CONS.CAN-NOT DE-TECT/REM.	
REFRIGERATED	0	+	0	+	0	+	III.
FROZEN	0	+	0	+	0	+	III.
CANNED	0	+	0	+	0	+	III.
(2) RAW MAT'S AND ING'S							
RAW CELERY	0	+	+	+	+	+	V.
DRIED MUSHROOMS	0	+	+	+	0	+	IV.
SOFT-RIPENED CHEESE		+	0	+	0	+	III.
SALT	0	+	0	+	0	+	III.
WATER	0	0	0	0	0	0	0.
STABILIZER	0	+	0	+	0	+	III.

NOTES: (1) AS USED BY CONSUMER
       (2) AS ENTERING THE FOOD FACILITY BEFORE PREPARATION OR PROCESSING

FIGURE 4-3.   Form 6.0-B. Risk assessment work sheet for physical food hazards, HACCP Principle 1.

Forms 5.0 (Microbiological, Fig. 4-1), 6.0-A (Chemical, Fig. 4-2), and 6.0-B (Physical, Fig. 4-3) illustrate the ranking of hazard characteristics and assignment of hazard categories for three modes of preservation for the cheese dip product, and the ranking of all raw materials and ingredients.

References

Corlett, D.A. and Stier, R.F. 1991. Risk assessment within the HACCP system. Food Control. 2:71–72.
Corlett, D.A. and Stier, R.F. 1990. Course manual: *A Practical Application of HACCP*. ESCAgenetics Corporation, San Carlos, CA. Licensed to D.A. Corlett, 5745 Amaranth Place, Concord, CA 94521.
National Advisory Committee on Microbiological Criteria for Foods (NACMCF). *HACCP Principles for Food Production*. FSIS Information Office, Washington, DC.

5

Determining Critical Control Points

William H. Sperber

Principle 2. *Determine the Critical Control Points required to control the identified hazards.*

INTRODUCTION

Rather than organize this chapter according to the types of hazards present in a food system—physical, chemical or biological—I have chosen to discuss Critical Control Points (CCP) according to a typical product flow from:

- production, growing or procurement of raw materials,
- ingredient receiving and handling,
- processing,
- packaging,
- distribution, and
- handling at retail, foodservice or in the home.

At each of these stages we will consider a number of CCPs for representative physical, chemical or biological hazards. Obviously, not every specific hazard and its CCP can be discussed in this chapter. However, an attempt is made to give a very comprehensive coverage of the possible CCPs.

Which type of hazard—physical, chemical, or biological—is the most commonly detected in food production? Physical hazards are the most common because of the many chances for foreign material contamination. Biological hazards, however, justifiably receive more attention because of the ability of microorganisms to multiply in food and potentially affect more people. For example, a stone or piece of glass in a package of vegetables may cause injury to a consumer, but it would affect only one individual and the injury would likely be minor. In contrast, *Salmonella* contamination in a pasteurized milk

39

packaging operation could affect many thousands of consumers and some of the resulting illnesses could lead to death.

Remember, a CCP is defined as: any point or procedure in a specific food system where loss of control may result in an unacceptable health risk.

The hazard analysis and risk assessment process described in the preceding chapter requires the involvement of technical experts from a variety of disciplines including microbiology, toxicology, engineering, and regulatory compliance. It is unlikely that any one person could identify all of the potential hazards in a given food production system. Therefore, the same group of experts needs to be involved in the determination of CCPs since a CCP must be established for each identified hazard.

REPRESENTATIVE CCPs

Growing

All types of hazards—physical, chemical and biological—are potentially associated with the growing of animals and plants.

Often antibiotics are used to treat diseases in animals. Only approved antibiotics can be used and often they cannot be administered within a certain period before slaughter. This CCP is necessary to protect consumers who are sensitive to specific antibiotics and to reduce chances for evolution of antibiotic resistant pathogens in the human population.

The application of pesticides to crops is another CCP. Only approved pesticides can be applied and then in the amounts specified by regulation or by the manufacturer. The types of pesticides will vary with individual crops. The timing of application before harvest is important so that residues do not persist in the consumer product. Many processors will grow crops under contract with individual farmers and closely regulate the use of pesticides.

Even the location of the growing field is a CCP. It is important to know the pesticide history of the field. Only approved irrigation water and systems should be used. For example, trench irrigation may be more appropriate than spray irrigation in some situations. Some processors require that fields not be located next to roads or former dump sites in an effort to minimize glass contamination. Fields should be carefully inspected before planting and objects such as glass bottles must be removed.

Ingredient receiving

Food ingredients should be shipped only in vehicles which are clean and sanitary. Nonfood chemicals, such as pesticides, cannot be permitted in the same shipment with food-grade materials. Bulk shipments in particular must be locked and sealed to assure that contamination or tampering cannot occur during shipment.

Temperature control of perishable raw materials is essential and will be a CCP.

Sensitive ingredients must be quarantined and tested before being released to production. Often the tests are performed before the shipment is unloaded (e.g., antibiotic or phosphatase testing in milk or aflatoxin testing in corn). Many of the microbiology tests require one or more days before completion. The ingredients must either be quarantined during this time or pre-shipment arrangements need to have been made with the supplier to provide the necessary assurance that the material is contaminant free. We, as other major food processors, maintain sensitive ingredient categories for *Salmonella, Staphylococcus aureus, Listeria monocytogenes, Bacillus cereus,* and aflatoxin.

The growing importance of "Just-In-Time" (JIT) procurement and production places greater demands on the microbiology laboratory to provide accurate results on a timely basis. It is the responsibility of the food processor to certify the competency of the microbiology laboratory—either its own, or the contract laboratory it may use. Laboratory certification entails the certification of individual technicians through training programs and check samples, and periodic laboratory audits for use of approved testing procedures and compliance with good laboratory practices.

Ingredient handling

The food processor must establish and maintain CCPs for ingredient handling, both for bulk and packaged ingredients. Most of these CCPs are necessary to detect and contain potential physical hazards.

All bulk receiving lines should be locked to protect against accidental or premature unloading, tampering, and infestation. The outlet of each bulk system needs to be protected by a physical control device such as a sifter, magnet or filter.

Some claim that there should be only one CCP for a particular hazard in a given production system. For example, the CCP for metal contamination would be a metal detector on the finished packaged product.

However, I disagree with this approach since the basic premise of HACCP is to prevent hazards from occurring in the first place. The keys to HACCP are design and prevention. Therefore, a more pragmatic approach is warranted. It is much better to have physical devices upstream to detect foreign material contamination as early as possible. If a bulk ingredient is contaminated with metal, for example, it would be better to detect the contamination and isolate the ingredient before it could be used. In principle, you should not wait until a contaminated ingredient has been converted to a more expensive finished product before detecting the contamination.

Another advantage of early detection is the protection of processing equipment which often operates with high precision and usually is quite expensive. Tramp metal which enters the processing stream either from an ingredient source or from a piece of equipment can be detected and immobilized by a number of devices such as magnets or screens.

At Pillsbury we require at least one CCP on each bulk ingredient system which feeds into a production line (Fig. 5-1A). In this way, problems can be identified and corrected more quickly than if we had relied solely on downstream CCPs. Bulk liquid systems such as liquid shortening or liquid sugar can be protected by filters or screens. Bulk dry systems such as bulk flour or bulk sugar can be protected by magnets, sifters or screens.

Packaged ingredients must be passed through physical devices which serve as critical control points. Bagged dry ingredients such as flour, nonfat dry milk, soy flour, and cocoa need to be emptied into a bag dump station which at a minimum includes a scalping screen and a magnetic grate. Optimally the bag dump station will also include a sifter with magnetic metal or synthetic Nytex screens. Canned ingredients such as No. 10 cans of mushrooms need to be dumped over a plate magnet to detect and remove metal shavings which inevitably

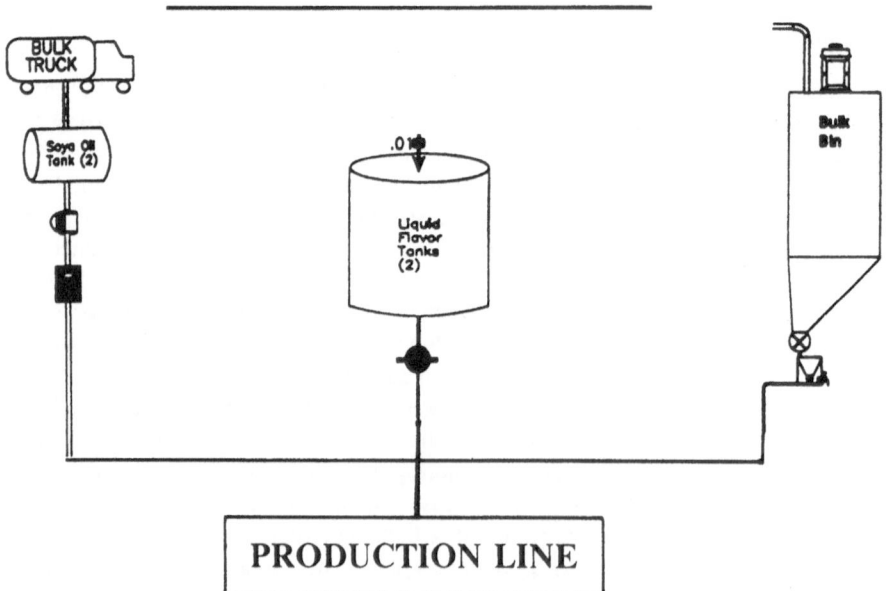

FIGURE 5-1.   CCP for each ingredient delivery system.

are formed when the cans are opened. Boxed ingredients such as nutmeats which will be added directly to products like ice cream should be passed through a metal detector before the nutmeats are added to the production stream.

Processing

The CCPs most obvious to us are those involved in food processing: the mixing of ingredients, often a thermal processing step, and the packaging of the completed product.

Some perishable foods depend on the product's formulation to assure their safety. In these foods, CCPs are necessary to control parameters such as the product's pH, water activity, or the presence of preservatives such as sodium nitrite.

It is common practice to save partial batches of product, as well as recycled or mislabelled products for later reworking. Usually the salvaged product is reworked into the new product stream at the rate of 5 to 10 percent. If any of the salvaged materials contain allergenic ingredients such as peanuts, milk, eggs, etc., a CCP must be established for rework control. This is essential to protect those consumers who are allergic or hypersensitive to these ingredients. A typical requirement at this CCP is to rework "like into like" so that cross-contamination of products which do not contain allergic ingredients cannot occur.

There are a great many types of thermal processes which can be controlled to assure the destruction of pathogenic microorganisms. Some of the processes provide a pasteurization treatment in which vegetative cells are killed, while others provide a sterilization treatment which is capable of inactivating the heat resistant bacterial spores.

The cooking of raw meats, poultry, and other ingredients can be established as a CCP for the destruction of vegetative pathogens such as *Salmonella* sp.

For some foods, the thermal process must be very precisely controlled to ensure the destruction of vegetative pathogens without overheating the product. Such precise controls are necessary for the pasteurization of milk, ice cream and liquid eggs. The equipment and controls used in these pasteurization processes are quite elaborate and are designed according to the published requirements of government regulatory agencies to assure the control of both temperature and time. Critical control points are established for indicating thermometers, recording thermometers and charts, differential pressure controls and flow diversion valves.

Some products depend on a hot-filling process to pasteurize both the product and its container. For these products, the temperature at filling is a CCP. The temperature needs to be maintained for a minimum time and with some containers, particularly plastic bottles, inversion is required to pasteurize the bottle's neck and spout.

Pancake syrups are a good example of a hot-filled product which requires CCPs for filling temperature, minimum time in the bottle above this temperature, and inversion of the bottles during their pasteurizing cycle. (Upstream, the pancake syrup requires formulation CCPs for the control of pH, water activity, and the presence of preservatives.)

The CCPs involved in sterilizing processes are analogous to those used in pasteurizing processes except there are more CCPs in sterilization. This is due to the physical complexity of the system required to achieve and maintain precise time and temperature control at high steam pressures.

Additional time and temperature controls are frequently the subject of CCPs. These include holding kettles, recycle systems, surge hoppers for fillers, cooling rates and storage temperatures. These CCPs are very important in the production of refrigerated and frozen foods.

In the above discussion we have been considering heat treatments of moist foods. Similar CCPs are necessary to control the pasteurization of dry food: e.g., "hot-boxing" egg albumen to assure the destruction of salmonellae.

The remaining processing CCPs to consider are those involved with facility design, good manufacturing practices (GMPs), and cleaning and sanitizing practices. Many of these CCPs are new to the food industry and have been necessitated by the emergence of *Listeria monocytogenes* as a foodborne health and regulatory hazard.

Facilities must be designed (or remodeled, if necessary) to maintain positive air pressure in the processing and packaging areas. This is a CCP which will keep airborne contaminants out of the production areas. The make-up air for the facility must be filtered and all areas around the intakes need to be kept clean and dry.

Similarly, process air needs to be filtered so that it cannot serve as a source of contamination. A good example of this is the air used to produce overrun in ice cream production. The presence of a clean and dry filter in this air line is a CCP.

We have strengthened a number of GMPs to control cross contamination from raw materials to processed foods. These are easily administered as CCPs. Examples include traffic control and some of the personnel GMPs. Traffic control applies to all persons—line workers, supervisors, visitors, maintenance workers and truck drivers. Under no circumstances should truck drivers be permitted inside the plant. Each plant should provide a separate waiting area for them. Line workers who handle raw material (e.g., raw meat) should not be permitted to handle cooked products or even enter the processing and packaging areas for the cooked products. The use of different colored uniforms and physical barriers can help enforce traffic controls. These controls also need to be applied to fork lift trucks and pallets. Some companies take traffic control so seriously that they have built separate locker rooms and lunch rooms for workers who handle raw materials and those who handle finished products.

The personnel GMPs which need to be enforced as CCPs include: hand dip stations in processing areas, footbaths with sanitizers and the wearing of hard-soled shoes or rubber boots instead of soft-soled (running) shoes which act as sponges to incubate bacteria and carry them from one area to another. As you can probably recognize, we wouldn't have talked about those practices as CCPs before the emergence of *Listeria*.

Similarly, certain cleaning and sanitizing procedures are now administered as CCPs. These are principally in post-pasteurization areas before products are packaged. We have established CCPs for cleaning and sanitation of certain food contact surfaces (e.g. ice cream freezers and fillers). I would expect that such CCPs could also be applied beneficially in the packaging of refrigerated meats.

Equipment cleaning can also be a CCP to assure the removal of allergenic ingredients before product changeovers.

Packaging

Several important CCPs are usually established in conjunction with product packaging. One of these is the use of a metal detector to reject products which contain metal. Another CCP is the use of a product code to provide trace and recall capabilities. A relatively new packaging CCP is the use of a tamper-evidence feature such as sealed membranes or shrink bands to protect consumers against product tampering.

A number of potentially important packaging CCPs must be considered and incorporated as necessary during the stage of product and package design. These include compliance with the consumer product safety act so that, for example, packages will not be set on fire in microwave ovens. Others are: labelling, so that consumers are aware of the presence of potential allergens; recipe review and approval, so that the consumer is advised not to use the product in an unsafe manner; the potential for the migration of chemicals from the packaging material to the food; and the composition of the headspace atmosphere. This last point is an important consideration for certain perishable foods where an anaerobic environment may permit growth and toxin production by *Clostridium botulinum*.

Distribution

The important CCPs in distribution are required for time and temperature control. Refrigerated foods must be kept at 40°F or below and frozen foods at or below 0°F. It is essential for the manufacturer to adequately chill or freeze the products before they are loaded onto trucks for distribution, since the refrigeration systems on trucks are designed to maintain temperatures, but cannot lower temperatures. It is also essential that products not be exposed to higher temperatures on loading docks for too long a period of time. The CCPs for distribution are particularly important in the United States and Canada because of the very long supply lines, in contrast to European countries where the supply lines are much shorter.

Retail/Foodservice/Home

Time and temperature controls are equally important during the final stages of getting the products to the consumer and during their storage in the home. Because of the unreliability of temperature control in many retail display cases, several manufacturers of chilled convenience foods have purchased refrigerated kiosks in which they themselves can control the temperature at the retail outlet. Foodservice applications extend over a broad range of settings including super-market delicatessens, restaurants, hotels, and institutions. In foodservice, CCPs for temperature control also apply to the proper holding of heated foods at elevated temperatures so that bacterial pathogens cannot multiply. This is a most important point, since over the years, inadequate holding temperature has been the most frequent cause of foodborne illness in foodservice operations.

In the home and in foodservice operations cross-contamination control is an important CCP to prevent the transfer of pathogens to the cooked food from raw foods, the utensils used to handle these foods, and the foodhandler's hands. Cooking procedures are also important CCPs, especially in the case of items such as raw poultry which are frequently contaminated with bacterial pathogens.

LOW-ACID CANNED FOODS CCPs

An example of determining CCPs can be given by reviewing those important in the production of a low-acid canned food (LACF) such as cream-style corn. It is most appropriate to use a LACF as an example because in 1974, the FDA incorporated the idea of HACCP and CCPs into a regulation involving the production of food for the first time, the LACF regulation (21 CFR, part 113).

The CCPs for LACF production can be organized into four groups: growing, processing and can filling, thermal processing, and can seam evaluation. There are many variables involved in thermal processing, each of which needs to be managed as a CCP. At first glance, some of the CCPs may not seem directly important to food safety. In fact, deviations at some of them may not lead to a health hazard. However, because thermal processing is a very complex and integrated operation, it is important that all CCPs be managed closely to assure food safety.

Growing

The CCPs employed here are the same as those we discussed at the outset and are based upon the selection of the growing field and the proper use of pesticides.

Process and can filling

The corn is removed from the cob and cut by rotating knives. One CCP would be the use of a magnet or metal detector to detect the break up of any of the knives.

Product viscosity is an important CCP because it has a major effect on the rate of heat penetration in either still or agitating retorts.

The amount of headspace is crucial in cans which will be thermally processed in an agitating retort. The initial headspace provides a bubble of air which serves to stir the can's contents as it rotates in the retort. If the bubble is too small, the rate of heat penetration will be reduced.

Can codes must be used for trace and recall capability. Sanitation surveys and bacterial spore counts on certain ingredients (e.g., sugar and starch) are important to manage the microbial load which the product carries into the thermal process. If the load is too high, commercial sterility might not be achieved.

Thermal processing

There are many CCPs involved in the actual delivery of the thermal process. Most of them are related to the equipment and can be observed in Fig. 5-2, which depicts a horizontal still retort. The CCPs at this stage are:

- Control of the initial temperature (IT) of the canned product as it enters the retort.
- Proper venting of the retort to assure that all air is removed as it is filled with steam.
- The thermal process must be conducted for a given time and temperature to provide commercial sterility. In the case of LACFs, the attainment of commercial sterility also guarantees the attainment of product safety.
- A mercury-in-glass thermometer is required to indicate the cooker temperature.
- Recording thermometers are required to document both time and temperature control.
- Sanitizers must be used in the cooling water so that those cans with microleaks in the seam area will not become contaminated.
- In the case of still retorts a system of crate control is an essential CCP to guarantee that unprocessed containers cannot bypass the retort.

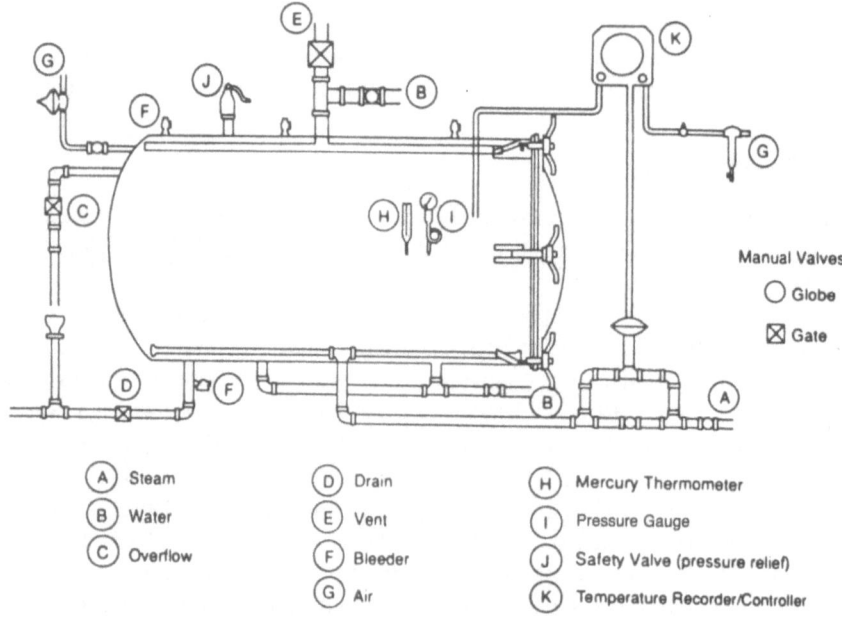

FIGURE 5-2.   Horizontal still retort.

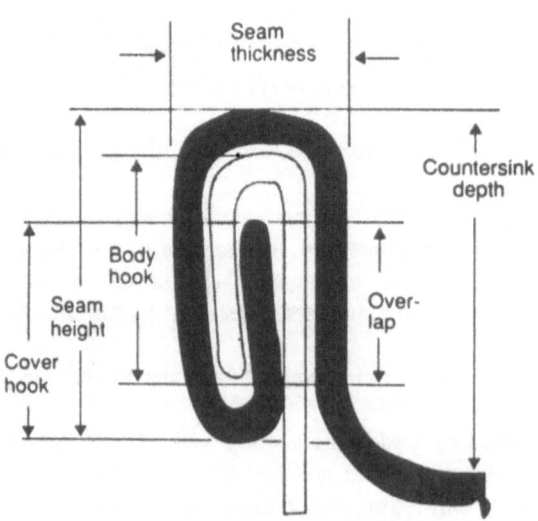

FIGURE 5-3.   Can seam evaluation parameters.

Can seam evaluation

The integrity of the can seams is vital to maintain product safety and stability during the relatively long shelf life of canned foods. Accordingly, there are many CCPs for can seam evaluation to assure the proper functions of the can seaming equipment. These include a visual examination for rough or sharp edges and other obvious defects.

Additionally, precise external dimensions are recorded (Fig. 5-3). These are: seam height, seam thickness, and countersink depth. At this point the can is cut apart and several internal dimensions are measured and recorded: cover hook length, body hook length, and the overlap between the two. Additional measurements not depicted in Fig. 5-3 are the tightness (wrinkle) rating, the pressure ridge rating and the juncture rating (for three-piece cans).

SUMMARY

The determinations of CCPs is a complex and demanding process covering a very broad range of physical, chemical, and biological hazards in many different types of operations.

Major food processing facilities can be quite large with a number of bulk ingredient handling systems and rather long and intricate processing lines. Such systems are essential for high speed production and for maximum flexibility in product changeovers. In the author's experience, the potential food safety hazards present in a complex processing system can be pragmatically managed by the establishment of about 15 to 20 CCPs. Specialized systems such as LACF production have about twice as many CCPs.

This chapter was developed to give a broad idea of the representative CCPs which may be necessary in food used in any single processing system. Conversely, not all possible CCPs have been presented. The determination of CCPs for a given production system depends on a complete and accurate hazard analysis by a team of experts who understand the product and the process by which it is produced.

6

Establishing Critical Limits for Critical Control Points

Lloyd J. Moberg

Principle 3. *Establish the critical limits which must be met at each identified CCP.*

INTRODUCTION

To achieve maximum success in food product protection, HACCP programs should be restricted to safety. HACCP Critical Control Points (CCPs) should only be used to control those points in a food manufacturing process where lack of control will likely result in the development of a potential safety hazard. They should not be used to control nonhazardous situations. Too many control points to monitor, by inclusion of nonhazardous points, will dilute the focus on safety. With manufacturing resources to monitor HACCP CCPs already limited in most cases, inclusion of nonhazardous points will result in the personnel not understanding which are the truly critical points. The end result of such a disparate program will be that nothing is being adequately monitored. There will then be no assurance that the product being released meets all the safety requirements; a potential hazard may be delivered to consumers. Nonsafety related monitoring procedures should instead be part of a standard quality assurance program.

HACCP Principles No. 1 and No. 2 (NACMCF 1990) have been introduced and discussed in previous chapters. Examples of potential hazards of a chemical, physical and biological nature have been identified and their risks assessed. Flow diagrams have been generated for the entire production process, from raw material growth through consumer use. Critical Control Points to control the identified hazards have themselves been identified and placed on the flow diagram. The next part of the HACCP process focuses attention to establishing the Critical Limits for these Critical Control Points.

The questions now become "How does one identify the Critical Limits associated with the Critical Control Points? How are the parameters for Critical Limits established? How does one determine when a safety hazard may develop?"

The purpose of this chapter is to help answer these questions. The chapter is divided into two areas: general definitions and guidelines on establishing limits; and, examples of established limits for biological, chemical and physical Critical Control Points.

LIMITS: DEFINITION & GUIDELINES

Definition

Principle No. 3. Establish the critical limits which must be met at each identified CCP (NACMCF 1990).

Description. A critical limit is defined as one or more prescribed tolerances that must be met to insure that a CCP effectively controls a microbiological, chemical or physical health hazard. Critical Limits on these CCPs represent the boundaries for safety.

Guidelines

In determining the Critical Limit for each CCP, the first question that needs to be answered is:

1. What are the critical components associated with the CCP?

Critical components associated with a CCP are those factors critical to safety, where failure to provide sufficient control may result in a safety hazard. A Critical Control Point may have multiple factors or components which need to be controlled to assure product safety. For example, the thermal process for a canned food product has numerous factors, all of which are important to food safety (e.g., product consistency/type/viscosity, initial fill weight/temperature, etc.). Failure to control any of these variables may result in an underprocessed canned food, which would be a potential consumer health hazard. Examples of other CCPs and potential critical components are shown in Table 6-1.

Prior to determining Critical Limits, all components or factors associated with the CCPs must first be identified; the Critical Limits can then be established for each of these components. The Critical Limits would be determined by identifying the point or level at which the component would become a potential hazard. Critical Limits are the boundaries of safety for the factors associated with the CCPs. The limit may be in various forms: a minimum holding temperature for a heated product, a maximum pH for an acidified beverage, an initial fill weight for a canned retorted product, a maximum holding temperature and time for a refrigerated, ready-to-eat product, etc.

In one example, a canned food product would have a recommended thermal

TABLE 6-1   Critical Components of Critical Control Points

Critical Control Point	Critical Components[a]
Retort process of canned food	Initial temperature
	Fill weights of can
	Pressure of retort
	Time at retort temperature
	Retort temperature
Heat process of RTE meat pattie	Oven temperature
	Time at temperature
	Pattie thickness
Metal detection on cereal line	Calibration metal detector
	Sensitivity/Capability
Acid addition to acidified beverage	pH of finished product temperature

[a]Examples of critical components are not inclusive of all factors important for the CCPs.

process from a process authority that specifies a minimum cook to assure safety. The thermal process is identified as a Critical Control Point in a HACCP program, since loss of control at this point (i.e. improper thermal process) would likely result in the development of a health hazard (i.e. *Clostridium botulinum* toxin, botulism). Critical components of this CCP would be those factors that have an influence on the proper thermal processing of this product. Some of the critical components are shown in Table 6-1. Associated with each of these critical components is a limit, exceeding which would allow a potential health hazard to develop. In most instances of retorted products, process authorities can identify the specific limits for each component. With new canned products for which the thermal process authority has no process, experimental testing is done to provide the necessary data from which limit information can be calculated.

In another example, an acidified beverage may have the acid addition to the batch process specified as a Critical Control Point. Failure to add the acid at this step would result in a final product pH that would be unacceptably high. Since the thermal process of the beverage (hot fill-hold) is determined by this pH, an elevated pH may result in an underprocessed product, or a product that would support the growth of pathogenic sporeforming bacteria, again a potential health hazard. The Critical Limit at this CCP would be pH 4.6. Exceeding a pH of 4.6 would allow the outgrowth of sporeforming pathogens if present. Note that the Critical Limit is not necessarily the target pH at which the product was intended to be produced (i.e. pH 3.8), but rather the pH at which the product could become a health hazard. Exceeding the target pH of 3.8 may result in an unacceptable product from a quality standpoint, but as long as pH 4.6 were not exceeded, a health hazard could not develop.

With the diversity of CCP factors that may affect safety, many different types

**TABLE 6-2   Criteria Most
Frequently Utilized for
Critical Limits**

Time
Temperature
Humidity
Moisture level (a_w)
pH
Titratable acidity
Preservatives
Salt concentration
Available chlorine
Viscosity

of limit information may be needed for safe control of a CCP. Examples of the different measures or types of information frequently utilized for Critical Limits are shown in Table 6-2. Whereas each CCP is important to the safety of the product being manufactured, the multiple factors associated with the CCPs are also important, as are each of the Critical Limits associated with these factors. If any one of the Critical Limits is out of control during the process, the CCP will be out of control and a potential hazard may then exist or develop.

Various resources will probably be used to identify the critical components of CCPs. These same resources can provide information on product safety for specific products and critical control points. These resources may include, but not be limited to, HACCP experts or recognized authorities, thermal process

TABLE 6-3   Sources of Information on Critical Limits

General Source	Examples
Surveys	Literature searches
	Supplier records/data
Regulatory guidelines	USDA requirements
	FDA regulations
	ICMSF, Codex, NAS/NRC
Experimental studies	In-house experiments
	Competitive microbes
	Inoculated packs
Experts	Thermal process authorities
	Consultants
	Microbiologists
	Equipment manufacturers
	Sanitarians

authorities, consultants familiar with the process, engineers, microbiologists, sanitarians, and common sense. Examples of sources of information are shown in Table 6-3.

After identifying the factors or components for each CCP, the second question that needs to be answered in order to identify Critical Limits is:

2. At what point or level would each of these critical components become a safety hazard?

To answer this question, information related to product safety must be available. As discussed earlier, various sources of information may provide information on critical components of CCPs. These same sources can be used to help determine the Critical Limits (Table 6-3). No information or lack of specific information for the product from these sources will require the manufacturer to experimentally determine and confirm the proper Critical Limit. This is also true if the capability of the manufacturing equipment to produce within specified parameters is unknown.

When the Critical Control Point is associated with manufacturing equipment, the processing variation associated with the equipment needs to be determined when establishing the Critical Limit. As an example, if cooking temperature were critical to the safety of a product, the target temperature of the oven would need to take into account oven temperature variation to assure adequate cooking of the product. Thus, a product requiring a minimum temperature of 150°F could not be baked in an oven programmed to bake at 150°F with a temperature variation of ±5°F; 50% of the product would be undercooked and potentially dangerous. The true temperature would need to be determined by measuring the actual temperature variation of the oven, and then selecting a target temperature that would assure no product would receive less than the 150°F cook. Ideally, this variation should be determined in a plant test before full manufacturing production occurs.

While the above represents a simple example, and other factors would also need to be considered, it emphasizes the point that equipment variation will influence Critical Limits. By identifying the variation associated with the equipment, the proper manufacturing target point can be set to assure continuous processing within the critical limit parameters. If the Critical Limit is exceeded during the manufacturing process, the Critical Control Point is out-of-control; a potential health hazard now exists. Any product produced while the CCP is out-of-control would need to be held and evaluated for safety. The out-of-control situation would need to be corrected prior to more product being produced.

The selection of specific parameters for a Critical Limit must be based on sound reasoning. The decision criteria on the selection of a Critical Limit should be based on the following considerations:

Exceeding the Critical Limit indicates: (Corlett 1988)

- Evidence of the existence of a health hazard (e.g., hazardous metal findings on the final magnet)
- Evidence that a health hazard could develop (e.g., underprocessing of low-acid canned food)
- Indications that a product was not produced under conditions assuring safety (e.g., metal detector kick-outs)
- Indications that a raw material may affect the safety of the product (e.g., pesticide audit detects aldicarb at high levels)

The most desirable characteristic of a Critical Limit is that it be easily measurable. In most cases, measurements can be achieved through use of automated instruments. Such automated devices, if used in the process flow, would ensure monitoring of 100% of the production. For a HACCP program, 100% monitoring of Critical Control Points assures the manufacturer that all product was within the specified limits for safety. Therefore, all product released for commercial distribution is safe for consumers.

ESTABLISHING LIMITS FOR CRITICAL CONTROL POINTS

Microbiological limits

Some Critical Control Points have identified microbiological hazards of bacterial, viral or parasitic nature which need to be monitored and controlled to assure product safety. Procedures at these CCPs would be designed to (1) destroy, eliminate or reduce microbiological contamination, (2) prevent recontamination, and (3) inhibit growth and/or toxin production. To accomplish these goals, the manufacturer may rely on utilizing physical processing systems as well as utilizing intrinsic characteristics of the food or the addition of salt or other preservatives.

To assure product safety, microbiological control at these CCPs will need to be monitored and verified. However, microbiological testing is not the method of choice for controlling CCPs. Microbiological testing is seldom effective for monitoring CCPs or their Critical Limits due to the time-consuming nature of the testing. Rapid microbiological screening tests for pathogenic microorganisms generally take 48 hrs to determine potential "positive samples". Confirmation of these positive samples may take several more days. It is generally not feasible to hold 2–4 days of production, especially limited shelf life products, while awaiting the results of microbiological testing.

Microbiological testing for CCPs is ineffective for yet another reason. Most product contamination by pathogenic microoogranisms will probably be at a low level (<1%). The probability of detecting a microbiological pathogen that is contaminating a product as such a low level is itself extremely low. Statistical sampling of the lot verifies that the chance of failing to detect the pathogen is significantly greater than its detection (Corlett 1991). The International Commission on Microbiological Specifications for Foods (ICMSF 1986) in their 2-Class or 3-Class Attribute sampling plans place the probability of accepting a product lot that contains a microbiological defect (defect rate of 0.1%) at 0.942 to 0.999. This means that there is a 0.1% to 5.8% chance of detecting the microbiological pathogen. The higher rate of detection is when 60 samples of the suspect lot are tested. The cost to test 60 samples per lot, coupled with a 5.8% chance of detecting the pathogen (0.1% defect rate), makes microbiological testing both cost prohibitive and ineffective.

In place of the time-consuming, cost prohibitive, and ineffective microbiological testing, physical and chemical measurements can be used as indirect measures of microbiological control. In these instances, the correlation between the physical or chemical parameters with the microbiological parameters would first need to be determined. With this correlation, exceeding the physical or chemical limit would mean the corresponding microbiological limit also would have been violated; a potential health hazard may then exist or develop. Thus, once correlated, instead of measuring the microbiological sterility of a canned food product after thermal processing, the physical measurement (e.g., retort time/temperature) would be used to indicate whether a microbiological problem may exist; an adequate thermal process would indicate all dangerous microorganisms had been destroyed. Instead of measuring the microbiological sterility of an acidified beverage, the pH of the product as well as the thermal process would be monitored to assure proper acidification and heating. An effective HACCP program will use continuous monitoring of physical (e.g., time and temperature parameters, etc.) and/or chemical (e.g., pH, titratable acidity, etc.) measurements to provide such assurance.

Microbiological testing may need to be done initially to identify the microbiological safety limits in the manufacturing process. This may be particularly true with new, innovative products for which safety information is nonexistent. In many instance, experimental studies (e.g., inoculated packs, challenge tests, etc.) will be used to determine the parameters (e.g., time/temperature, pH limits, etc.) at which the microorganisms reach a health hazard level, or to verify that the Critical Limits are adequate to control the hazardous microorganisms. Once established, these microbiological limits can be correlated with physical or chemical limits; these physical and chemical limits will then serve as indirect measurements of microbiological control during the normal manufacturing process. Microbiological testing can also be used to spot check (i.e. audit) that the identified microbiological safety limits remain valid.

Some manufacturers regard microbiological specifications on incoming ingredients as a Critical Control Point in their HACCP program, and thus require a Certificate of Guarantee from the supplier. This may be particularly important when (1) there is a reasonable probability that the raw material may contain microbial pathogens, (2) the supplier's process contains a destruction step to eliminate the pathogens, and (3) neither the manufacturing process nor the consumer preparation process contains a "kill" or destruction step to destroy microbial pathogens. Incoming ingredients would therefore need to be "pathogen-free." For such "Certificates of Guarantee" to be meaningful, the ingredient supplier would need to manufacture using a functioning HACCP program. Suppliers who utilize HACCP will be able to assure compliance with the microbiological specifications due to proper chemical and physical control and limits within their processing systems. The alternative approach of using end product testing to assure microbiological safety provides little assurance that the ingredient is safe. As noted earlier, the probability of detecting pathogens contaminating the ingredient at low levels using end product testing is extremely remote. The sources and procedures mentioned previously should be used to determine the microbiological limits for the ingredients.

To identify proper microbiological limits, the sources mentioned earlier should be consulted. Literature surveys or supplier records may contain the necessary information on pathogenic levels in an ingredient or product. Federal, State or Local government agencies (i.e. FDA or USDA) may have specific regulations on some foodborne pathogens (e.g., zero tolerance for *Salmonella* spp., *Listeria monocytogenes*). The National Academy of Science (NAS 1985) and ICMSF (ICMSF 1980; ICMSF 1988) publications can be used to identify acceptable/unacceptable levels of the microorganisms in a variety of food products. However, even with this information, the manufacturer may still need to do microbiological testing to establish or confirm the baseline levels in the products.

Chemical limits

Chemical hazards can be arbitrarily divided into two categories: naturally occurring and added. Some naturally occurring chemical hazards may have established maximum limits (e.g., aflatoxin, shellfish toxin, scombrotoxin). Other naturally occurring chemicals may come from toxic elements (e.g., lead, mercury) or toxic compounds (e.g., arsenic). Added chemicals may originate during the growth cycle of the raw ingredient (e.g., pesticides, insecticides, fungicides, herbicides, growth hormones, antibiotics, etc.). Other added chemicals may be inadvertently introduced during the manufacturing process (e.g., inks, lubricants, cleaners/sanitizers, etc.). While still other chemicals may fall into the category of food additives (e.g., vitamins, colors, preservatives, nitrites, sulfites, etc.). Regardless of the source, chemicals that may cause a food safety hazard fall into a HACCP program, and the appropriate Critical Control Points and their associated Critical Limits must be identified.

In determining the need to monitor for a particular chemical hazard, the manufacturer would base his judgement on the reasonable probability that such a chemical could be in his ingredient or product. Obviously, limited resources would make it fruitless to attempt to monitor for all chemicals without any rationalization that they may be in the product. Once the chemical hazards have been identified and the Critical Control Points determined, the Critical Limits can be set. As noted above, some limit information is already specified for naturally occurring chemical hazards.

Critical Limits for other chemical hazards would ideally utilize the level at which such a chemical would present a food safety hazard. However, if a chemical is not approved for the food, such as pesticides, it would be prudent for the manufacturer to set up a zero tolerance than determine safety level. Numerical limits would only be important when one formulates a product with specific chemicals (e.g., Vitamin A, FD&C Yellow No. 5, sulfites, etc.) and needs assurance that the associated safety limits are not exceeded. The potential inadvertent contamination of food products by nonhazardous adulterants (e.g., vegetable-based inks, food grade lubricants, etc.) which would not cause a health hazard would be better controlled by a Quality Control program rather than inclusion into the HACCP program. As mentioned earlier, inclusion of too many points to monitor, especially other quality and regulatory points, will dilute out the focus on safety.

Manufacturers who rely on supplier Certificates of Guarantee for their incom-

TABLE 6-4   Chemical Hazards and Associated Critical Limits

Chemical Hazard	Critical Limit
Agriculture chemicals (pesticides, fungicides, fertilizers, antibiotics, insecticides, herbicides)	Crop use and limits specified by EPA
Mycotoxins (including aflatoxin, ochratoxin, zearalenone, vomitoxin)	Aflatoxin (20 ppb) Other mycotoxins have no identified regulatory limit, or hazardous limit
Shellfish toxins	Undetectable by current methodology
Toxic elements and compounds (Pb, Hg, Zn, arsenic, cyanide)	As specified by Regulatory Authority
Processing plant chemicals (cleaning agents, lubricants, sanitizers)	Within regulatory limits or company policy; no visible adulteration of food
Food additives (preservatives, flavor enhancers, nutritional additives)	Specified by FDA (Direct food additives)

ing ingredients must assure themselves that monitoring was done via a HACCP program rather than end product testing. Like microbiological Critical Limits, chemical Critical Limits that rely on attribute sampling cannot provide high assurance that defects will be detected. Ideally, continuous monitoring of CCPs for their associated chemical factors will provide the best assurance of conformance to safety standards.

Identification of chemical Critical Limits can utilize the sources identified in Table 6-3. Additionally, experts for chemical hazards and limits may also include toxicologists and plant pathologists.

Examples of chemical factors that may be important to control and possible critical Limits are presented in Table 6-4.

Physical limits

Identification of physical hazards in a food processing system is straightforward. Any physical matter that is not normally found in a food is considered an adulterant. Those that present a health hazard are physical hazards of concern. Such physical hazards would include: glass, metal, wood, stones, bones, plastic, and employee personal effects. To be hazardous, these materials would be of such size and shape that they would pose a potential health hazard concern.

Limits on CCPs associated with physical hazards are the most straightforward. Critical Limits on physical hazards will be zero or nondetectable. Metal detectors, magnets, screens and sifters can be used to detect most physical hazards. Functioning of this equipment would be such that physical hazards would be removed or detected to meet the zero or nondetectable Critical Limit. The most important function in this area is not in setting the limit, but in assuring proper installation of the equipment in the system, in verifying the calibration of the equipment, in checking of tailings (screens and sifters) for the foreign material, and in maintaining the equipment.

Most manufacturers specify on their purchase orders or their Ingredient Specifications that all product delivered from a supplier must be free of "all forms of foreign and extraneous matter as can be achieved by Good Manufacturing Practices". The manufacturer should also verify the existence of a supplier HACCP program, or at a minimum, sufficient product protection devices are present, functioning, and properly maintained to prevent physical hazards from contaminating his potential ingredients.

Sources that may be useful in identifying proper equipment for physical hazard detection or removal would include the equipment manufacturers, engineers, and appropriate texts (Imholte 1984). Once the proper equipment is sourced and installed, proper calibration and maintenance is essential, as well as accurate and complete documentation of its operation and findings.

TABLE 6-5   Physical Critical Control Points and Example Control Limits

Physical CCP	Critical Limit
Metal detector	• Rejection of 3/32″ Series 400 stainless steel sphere 100% of the time (calibration) • No hazardous metal, ferrous and nonferrous detectable
Magnet	• No hazardous metal
Screen	• Size (will depend on product) • In good repair • Tailings (no hazardous findings, specific requirements are product dependent)
Sifter	• Size (will depend on product) • In good repair • Tailings (no hazardous findings, specific requirements are product dependent

Examples of physical hazard Critical Control Points and their associated Critical Limits are presented in Table 6-5.

CONCLUSION

Limits on microbiological, chemical and physical hazards represent the boundaries of safety for the Critical Control Points. Exceeding these boundaries means that a health hazard may exist or could develop, that the product was not produced under conditions assuring safety, or that a raw material may affect the safety of a product.

Microbiological testing is ineffective due to inefficiencies in time and detection methodology. Chemical and Physical monitoring can be used to indicate microbiological process control, in addition to chemical and physical control. Ingredient suppliers who utilize "Certificates of Guarantee" should have a functioning HACCP program to verify absence of health hazards and should not rely on end product testing.

Resources available to help determine the critical components of CCPs as well as their associated limits were identified. To maximize food product protection, Critical Control Points and their Critical Limits should only be used to control those points in a process where lack of control will likely result in the development of potential safety hazards.

References

Corlett, D.A. 1988. Monitoring and decision criteria. In *The HACCP Manual* for The Food Processors Institute course on "Workshop on Establishing Hazard Analysis Critical Control Point (HACCP) Programs." The Food Processors Institute, Washington, DC.

Corlett, D.A. 1991. Monitoring a HACCP System. *Cereal Foods World* 36: 33, 34, 36, 38, 40.

Imholte, T.J. 1984. *Engineering for Food Safety and Sanitation. A Guide to the Sanitary Design of Food Plants and Food Plant Equipment.* The Technical Institute of Food Safety, Crystal, MN.

International Commission on Microbiological Specifications for Foods (ICMSF). 1980 *Microbial Ecology of Foods. Vol 2. Food Commodities.* Academic Press, New York.

International Commission on Microbiological Specifications for Foods (ICMSF). 1986. *Microorganisms in Foods.——Sampling for Microbiological Analysis: Principles and Specific Application,* 2nd ed., ICMSF. Univ. Toronto Press, Toronto.

International Commission on Microbiological Specifications for Foods (ICMSF). 1988. *HACCP in Microbiological Safety and Quality.* Blackwell Scientific Publications, Oxford.

National Academy of Sciences (NAS). 1985. *An Evaluation of the Role of Microbiological Criteria for Foods and Food Ingredients.* Food Protection Committee, Subcommittee on Microbiological Criteria. National Academy Press, Washington, DC.

National Advisory Committee on Microbiological Criteria for Foods (NACMCF). 1990. *HACCP Principles for Food Production.* USDA-FSIS Information Office, Washington, DC.

7

Monitoring Critical Control Point Critical Limits

Martha Hudak-Roos and E. Spencer Garrett

Principle 4. *Establish procedures to monitor critical limits.*

The activity of monitoring within a HACCP system is essential to the system's success. In order to establish and effectively conduct monitoring procedures, the questions what, why, how, where, who and when must be answered.

WHAT IS MONITORING?

Monitoring, as defined by Webster, is "watching, observing, or checking especially for a special purpose." Within a HACCP system, monitoring has been defined as: (1) "Checking that the processing or handling procedure at a CCP (Critical Control Point) is under control" (ICMSF 1988); (2) "The scheduled testing or observation of the effectiveness of a process to control 'Critical Control Point(s)' and their limits" (FDA/NOAA 1990); and (3) "A planned sequence of observations or measurements of critical limits designed to produce an accurate record and intended to insure that the critical limit maintains product safety" (NACMCF 1990).

All of these definitions agree that monitoring is an action. While monitoring may be done by a continuous instrument, it is not the same as continuing observation. Monitoring requires management action. It is not something that can be set up, turned on, and ignored.

WHY DO WE MONITOR?

Obviously, monitoring is done to collect data and subsequently have information upon which to base a decision. But monitoring also provides an early warning that a process is losing or out of control (Jarvis 1990). When done properly, monitoring can help to prevent or minimize loss of product when a process or handling deviation occurs. It can also help to pinpoint the cause of the problem

62

when control is lost. Without effective monitoring and recording of data or information, there is no HACCP system.

HOW DO WE MONITOR?

Monitoring can be done by observation or measurement (NAS 1985) at process or sanitation CCPs. In general, observations give qualitative indices and measurements result in quantitative indices. Thus, the choice of whether the monitoring will be an observation or measurement (or both) depends upon the established critical limit and available methods as well as realistic time delays and costs.

Since monitoring is a data collection activity, it is important to understand how to collect data. In general, there are ten steps to follow in designing a data collection (monitoring) activity (Anonymous 1990):

(1) Ask the right questions. The questions must relate to the specific information need. Otherwise, it is very easy to collect data that are incomplete or answer the wrong questions.
(2) Conduct appropriate data analysis. What analysis must be done to get from raw data collection to a comparison with the critical limit?
(3) Define "where" to collect.
(4) Select an unbiased collector.
(5) Understand the needs of the data collector, including special environment requirements, training, and experience.
(6) Design simple but effective data collection forms. Remember KISS—keep it simple, stupid! Check to see that the forms are self-explanatory, record all appropriate data, and reduce opportunity for error.
(7) Prepare instructions.
(8) Test the forms and instructions and revise as necessary.
(9) Train data collectors.
(10) Audit the collection process and validate the results. Management should sign all data forms after review.

Observation

Data collection by observation is the most basic. While monitoring by measurement often is recommended because it gives "unbiased" numbers, the importance of observations cannot be overlooked. For example, one of the best ways to collect data on the condition of your car's tires is to observe tread wear.

The NAS gave examples of monitoring observations for food service in their 1985 *An Evaluation of the Role of Microbiological Criteria for Foods and Food Ingredients* (Table 7-1).

TABLE 7-1   Hazards, Critical Control Points, Preventive Measures, and Monitoring Procedures of Foodservice

Operational Step	Hazards	Critical Control Point	Preventive Measures	Monitoring Procedures
Storing incoming foods— Frozen storage	Thawing because of power failures or improper frozen food storage or transport.	Thawed food.	Keep frozen foods frozen; maintain product temperature at or below 7°C (45°F) after thawing.	OBSERVATION: See that foods are frozen and remain so.
	Prolonged storage.	Interval between freezing and use of food.	Rotate stock; use before detrimental effects occur.	MEASUREMENT: Measure temperature of freezing unit to determine whether it is −17°C (0°F); drill into frozen food and measure temperature to determine whether it is frozen.
	Foods show signs of spoilage	Usually thawed food.	Discard.	OBSERVATION: Compare date of processing, if known; expiration date, if known; or date of storage with date of use; look at texture of product. OBSERVATION: Observe condition of food for slime, mold, gas formation, off-odor, freezer burn, etc., that are characteristic of spoilage.

TABLE 7-2   Cooked Shrimp Process Steps and Control Points

Step	Hazard	Control Point	Preventive Measures	Monitoring
Receiving	Thermal abuse Decomposition Bisulfite Microbial pathogens Additives abuse a. bisulfite b. sodium hydroxide c. borates d. phosphates e. chlorine Contaminants—filth and extraneous materials Integrity of package Copacking problems (short weigh, dehydration, etc.)	Incoming raw materials Receiving room Unloading area	Purchasing specifications Control time/ temperature abuse Control Product Movement Adequate physical separation of raw from cooked shrimp	Temperature checks and visual observation Testing (microbiological, chemical, etc.) for specification compliance Sensory and visual examinations for decomposition

The NMFS, in their series of HACCP Regulatory Models for the seafood industry, also outlined examples of monitoring by observation (Table 7-2) (NMFS 1989).

As indicated in these tables, included in processing observations are sensory and visual checks for everything from decomposition to the location of a product in a chill room or refrigerator. Observation monitoring also is important for sanitation. Prior to process start-up, the most important sanitation monitoring tool is observation.

Of course, the observations must be compared to the CCP's critical limit(s). This requires a manual analysis by the observer and, in many cases, an interpretation or subjective call. Extreme care in selecting, training, and standardizing observers must be taken.

Observations generally are recorded on a "checklist." The checklist should contain items of importance or relevance to the specific CCP and its critical limit. A portion of a monitoring checklist is exemplified in Fig. 7-1.

Measurement

Monitoring by measurement can include physical, chemical, or microbiological indices. Measurement monitoring also can be used for both process and sanitation CCPs.

CHECKLIST EXAMPLE

SANITATION LOG				

Date:_____

S = Satisfactory
N = Needs Improvement
A = Alert

TIME:	Pre-Start	Break 1	Break 2	Comments
Thaw tank cleaned				
Glaze water changed				
Belts cleaned and in good repair				
Utensils cleaned and in good repair				
Processing machines cleaned				
Lighting				
Floor clean				
Ceilings without peeling paint or condensates				
Dip stations				
Trash removed				
Chlorine barrels				

Ins. By:_____   Prod. Supv.:_____   QA Mgr.:_____

FIGURE 7-1.

The most common process measurements taken are time, temperature, and pH. However, for raw materials, chemical tests for toxins, food additives, contaminants, etc., and microbiological tests for coliforms, *E. coli, Salmonella,* etc., often are used.

When not an observation, sanitation critical limits usually depend upon microbiological testing. Whether they be quick tests or standard methods, sanitation monitoring is designed to determine overall effectiveness of sanitation and not for the precise quantification of microorganisms (Nickelson 1978).

Measurement monitoring has taken on a degree of sophistication with the advent of the use of computers in food processing. Computers have been used to quantitate black spot in shrimp as well as clean-in-place sanitation systems (Larusson 1991; Eilers 1991).

As expected, measurement monitoring requires some extra care. Equipment must be calibrated and data collection must have quality control procedures. An uncalibrated thermometer or one that does not read to the desired decimal point can do more harm than good.

Measurement data can be collected a number of ways. The easiest way is a data sheet (Fig. 7-2). Data sheets should record data in a simple format. However, as in observation monitoring, care must be taken that the data collector is instructed sufficiently to perform the data analyses relative to the critical limit. In this example, the data collector must know that any internal temperature below 180°F is cause for a corrective action.

Data collection on check sheets or control charts, on the other hand, can be designed to more easily interpret results, demonstrate trends, and highlight subtle changes (Quality Progress 1990; Jarvis 1990; Nolan 1990). Not only are these techniques control devices, but action and analytical devices as well (Rosander 1985). By charting the data versus the critical limit, the data analyses is already on paper (Fig. 7-3 and 7-4). Further, for those who wish to examine variability or identify problem causes, these charts are a first step. Such an advanced use of these charts can be found in the Statistical Process Control/Statistical Quality Control (SPC/SQC) literature.

The trend for measurement monitoring, though, is toward full automation. Micro-processing systems can have visual and/or sound alarms when critical limits are defaulted. Automation can produce data sheets as well as control charts and check sheets. If calibrated and maintained correctly, automated systems can help to reduce the fear of human error.

DATA SHEET

I.T.	LINE 1	C.L. = 180°F
TIME	**TEMPERATURE**	
0800	181	
0830	181	
0900	180	
0930	180	
1000	179	
1030	179	

NOTES:

OPERATOR	DATE

FIGURE 7-2.

CHECK SHEET

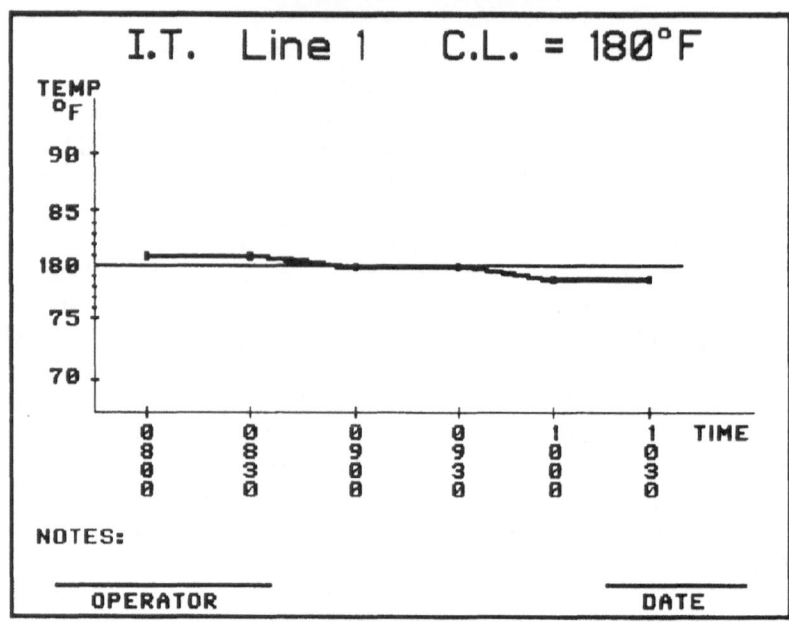

FIGURE 7-3.

CONTROL CHART

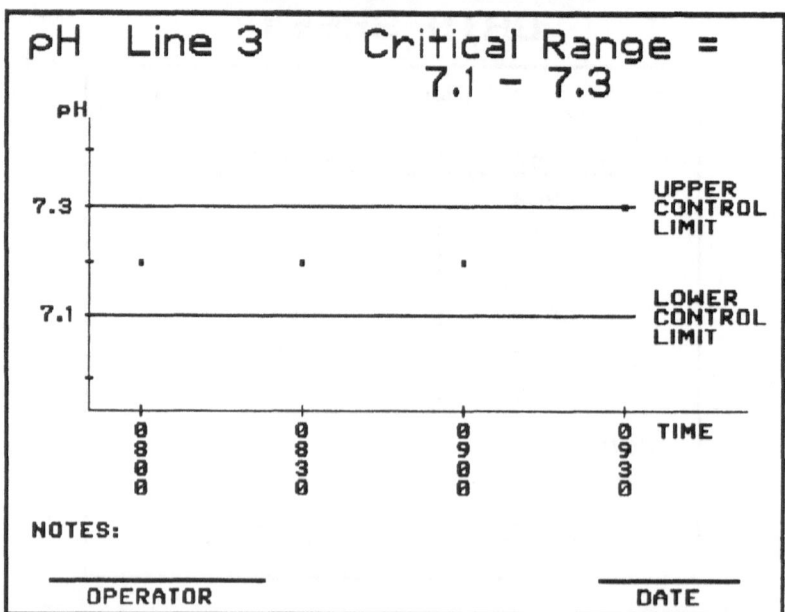

FIGURE 7-4.

WHERE DO WE MONITOR?

Monitoring in a HACCP system is done minimally at critical control points. It must be done at a location within a CCP that accurately reflects the state of the critical limit; however, the ideal is to monitor where there is minimal interruption in the production flow. For example, if the critical limit of a cooking CCP is an internal temperature of 180°F, then monitoring must be done during or post cooking when the maximum temperature has been shown to be reached.

Again, the key to establishing "where" is to learn to ask the right questions. Only then can effective data collection occur (Quality Progress 1990). Ask yourself:

- What questions need to be answered?
- How will the answers to this question be communicated?
- What data analysis must be done, and how will we communicate the results?
- What type of data do we need?
- Where in the process can we get these data?

Let's look at the internal temperature example above.

(1) What question needs to be answered?
    What is the internal temperature of the cooked product at the completion of the cook (or after a certain amount of time)?
(2) How will the answers to this question be communicated?
    The answer will be communicated by measuring the internal temperature.
(3) What data analysis must be done, and how will we communicate the results?
    The internal temperature must be compared to 180°F. We can communicate the results by an electronic alarm, by continuous monitoring of a temperature recorded, or by periodic examination of a temperature control chart.
(4) What type of data do we need?
    We need internal temperature data recorded to nearest 0.1°F.
(5) Where in the process can we get these data?
    We can get these data from the cooked product as it exits the cooker.

Some of these questions and answers are self-evident, and it might appear foolish to follow this process for every critical limit. However, the important concept is that, in deciding where to collect data, the process works backwards. Don't first ask yourself what data you need, but rather, what are the questions. This better defines the data needs and subsequently where the data should be collected.

WHO MONITORS?

The qualifications of the data collector must be based upon the how and where of monitoring. Certainly, the collector must have the easiest access to the CCP, and the skills and knowledge to understand not only the food production process but the purpose, importance, and process of the monitoring activity. In some cases (i.e. organoleptic determination of decomposition or chemical or micro-biological analyses), the person must have a high level of training and experience. Of course, the person should be unbiased.

All things considered, the "who" should be someone you can place your faith in.

WHEN DO WE MONITOR?

If monitoring is not continuous, the question of *when* becomes extremely important. It is no less important for in-process monitoring than for lot; intermittent or noncontinuous monitoring must reliably indicate that the hazard is under control.

Intermittent monitoring quickly leads to a discussion on statistics. If monitoring is on a per lot basis (i.e. raw material), the question becomes modified to "How much do I sample?" If intermittent monitoring on-line, then, in addition to how much, one asks "How often do I sample?"

These questions are best answered through statistical analysis. Once you or your firm's management has decided on the amount of risk you are willing to accept, then through the literature and/or competent statistical authorities your *when* questions can be answered.

SUMMARY

Monitoring is an action that requires management attention. Its purpose is to collect data for making a decision relative to the established critical limit. Monitoring is done at the point most relevant to the control objective. You can observe and/or measure when you monitor; all monitoring is recorded. Confidence in the person who has the monitoring responsibility is essential. Training the data collector and periodic audits of his or her performance are important. If monitoring is not continuous, then the amount of monitoring must be based upon the amount of risk that is acceptable to management. In order to make these decisions, competent statistical consultation is required.

References

Anonymous. 1990. The tools of quality. Part V. Check sheets. In *Quality Progress*. American Society for Quality Control, Milwaukee, WI. Oct.

Eilers, J. 1991. Computer monitoring yields optimum CIP system performance efficiency. *Food Processing* 52(2):

U.S. Food & Drug Administration, U.S. Dept. of Health & Human Services (FDA/NOAA). 1990. HACCP based plan submission guide. Washington, DC, Sept. 11.

International Commission on Microbiological Specifications for Food (ICMSF). 1988. *Microorganisms in Foods. 4. Application of the Hazard Analysis Critical Control Point (HACCP) System to Ensure Microbiological Safety and Quality*. Blackwell Scientific Publications, London.

Jarvis, G. 1990. HACCP, monitoring and statistics. In *Be Safe with HACCP—The Critical Component of a QA System*. B.C. Food Technology Centre, Industry Development Group, B.C. Research, Vancouver, British Columbia. August.

Larusson, T., Balaban, M., Otwell, S., and Yeralan, S. 1991. Application of computer vision to seafood quality evaluation. In *Proceedings of the Fifteenth Annual Tropical and Subtropical Fisheries Technological Conference of the Americas*. Sea Grant College Program, Univ. of Florida, Gainesville, FL. In press.

National Academy of Sciences (NAS). 1985. *An Evaluation of the Role of Microbiological Criteria for Foods and Food Ingredients*. National Academy Press, Washington, DC.

National Marine Fisheries Service (NMFS). 1989. HACCP regulatory model for cooked shrimp. Report of the Model Seafood Surveillance Project, National Oceanic and Atmospheric Administration, NMFS, Office of Trade and Industry Services, National Seafood Inspection Laboratory, Pascagoula, MS.

Nickelson, R. II. 1978. Food contact surfaces—indices of sanitation. In *Sanitation Notebook for the Seafood Industry*. VPI SG-78-05. U.S. Dept. of Commerce, National Sea Grant College Program, Washington, DC.

Nolan, K. 1990. Planning a control chart. In *Quality Progress*. American Society for Quality Control, Milwaukee, WI. Dec.

Rosander, A. 1985. *Application of Quality Control in the Service Industries*. American Society for Quality Control, Milwaukee, WI.

National Advisory Committee on Microbiological Criteria for Foods (NACMCF). 1990. *HACCP Principles for Food Production*. USDA-FSIS Information Office, Washington, DC.

8

Corrective Action Procedures for Deviations from the Critical Control Point Critical Limits

R. B. Tompkin

Principle 5. *Establish corrective action to be taken when there is a deviation identified by monitoring of a CCP.*

INTRODUCTION

The HACCP concept has continued to evolve so that today several variations of HACCP have been proposed. Do not wait for the final, official version to appear before implementing HACCP. Select the variation which best suits your company and begin to gain experience with the concept. The HACCP Plan should be incorporated into your operating instructions so it is a working document. The HACCP Plan should not be a separate book which is placed on a shelf and then forgotten until needed.

This chapter covers Principle No. 5, "Corrective Action". To review, HACCP is a system in which (1) the food and its intended use are described, (2) a flow diagram is prepared which describes the steps in the process which are under your control, (3) potential hazards are identified and prioritized according to risk and severity, (4) critical points in the operation are identified which permit control of the hazards, (5) criteria (e.g., time, temperature, pH) are specified which will provide control of the hazards and indicate whether the operation is under control, (6) rapid tests are used to monitor whether the CCPs are under control, and (7) corrective action is taken when monitoring results indicate the operation is not under control. Corrective action, then, is a response to monitoring.

CORRECTIVE ACTION ACTIVITIES

Corrective action involves four activites. The first activity is to use the results of monitoring to adjust the process to maintain control. Second, if control is

72

lost, you must deal with noncompliance product. Third, you must fix or correct the cause of noncompliance. Fourth, you must maintain records of the corrective actions which have occurred.

Defining control

The first activity is to adjust the process to maintain control. But, what is meant by control? The following is offered as a definition for control as used in a HACCP Plan: Control means "managing the conditions of an operation to maintain compliance with established criteria" (Tompkin 1990). This definition incorporates several principles of HACCP wherein criteria are established, the CCPs are monitored, and adjustment are made to maintain compliance with the criteria. This would appear to be a fairly straightforward process. A debate will often occur, however, when these principles are put into practice. The debate concerns whether the CCP is under control. Since corrective action depends upon the assessment of whether a CCP is under control, some further discussion of control and CCPs may be desirable.

A critical control point is an important element of a HACCP plan. The National Advisory Committee and the USDA defined a CCP as "Any point or procedure in a specific food system where loss of control may result in an unacceptable health risk" (NACMCF 1990). Another definition for critical control point may be "A step in a process at which control can be exercised and a food safety hazard can be minimized". The highly debatable issue of what is an acceptable or unacceptable health risk is omitted from the definition. The definition encourages the operator to strive for control even though it may be only possible to minimize a hazard. The definition clearly states that a CCP exists only when control can be exercised. It recognizes the reality that some hazards can be minimized but not prevented. This definition is derived from the ICMSF definition for a CCP (ICMSF 1988). It is also very similar to a definition for CCP which has been adopted by the Meat and Poultry Working Groups of the National Advisory Committee.

The characteristics of an ideal CCP are listed in Table 8-1. Unfortunately, we do not work in an ideal world an don't always have ideal CCPs. In many cases, it may be possible to minimize but not prevent a hazard. Also, the criteria cannot always be as clearly defined as we would prefer. Our assessment of whether the criteria are being met may be based upon the judgment and experience of the observer. In reality, our ability to prevent hazards ranges from partial to complete elimination of each hazard. This raises the question then, is the CCP under control? Are we managing the operation to maintain compliance with the established criteria? In the case of an ideal CCP as described in Table 8-1, the answer is very clear. The response is a yes or no. But, with a less than ideal CCP, the answer may be yes, no, or maybe, depending upon the circumstances

TABLE 8-1   The Ideal CCP

1. Criteria are supported by research and the technical literature.
2. Criteria are specific, quantifiable and provide a yes/no response.
3. The technology for controlling the CCP is readily available and at reasonable cost.
4. Monitoring is continuous and the operation is automatically adjusted to maintain control.
5. There is a favorable history of control.
6. The potential hazard is prevented or eliminated.

and who is doing the evaluation. One person will reach a decision, but if five different people are making the same assessment, there will likely be differences of opinion and a consensus will have to be reached in some manner.

Table 8-2 lists five examples of CCPs which can be controlled very effectively and which prevent a potential hazard. The first example is the pasteurization of milk for the destruction of nonsporeforming pathogens. The effectiveness of this CCP is supported by extensive literature and years of commercial experience demonstrating that, if the criteria for pasteurization are met, then nonsporeforming pathogens are destroyed. This includes *Listeria monocytogenes* which recently was suspected of surviving the pasteurization requirements. A second example is the use of proper containers for the storage of high acid foods and beverages. If proper containers such as stainless steel or glass are used for production and storage, metal poisoning can be prevented. If galvanized metal is used for acidic fruit juices, for example, then zinc would be leached from the galvanized metal with resulting poisoning of the consumers. The third example is refrigeration of foods at ≤10°C to prevent the growth of proteolytic *Clostridium botulinum*. There is extensive literature and commercial experience which demonstrates that, if foods are stored at <10°C, proteolytic botulinal outgrowth will not occur. A fourth example of a CCP which effectively controls the potential hazard is the acidification of foods such as canned vegetables or pickled sausages to a pH of 4.6 or below to prevent pathogen growth. The final example involves drying foods such as snack meats to a water activity of <0.86 to prevent bacterial pathogen growth. In each of these examples, there is a clearly defined answer

TABLE 8-2   Critical Control Point—Effective Control

1. Pasteurization of milk destroys nonsporeforming pathogens.
2. Using proper containers for high acid foods and beverages prevents metal poisoning.
3. Refrigeration at 10°C (50°F) or below prevents the growth of proteolytic *C. botulinum* in food.
4. Acidifying canned foods to pH 4.6 or below prevents pathogen growth.
5. Drying foods to a water activity of 0.86 or below prevents bacterial pathogen growth.

TABLE 8-3   Critical Control Points—Partial Control

1. Bone, metal, and stone collectors reduce the risk of hard foreign objects in food.
2. Sorting peanuts with properly adjusted and maintained equipment will minimize the presence of mycotoxin in peanut products.
3. Careful evisceration reduces contamination of raw meat and poultry with enteric pathogens.
4. Training and reminding employees about proper hygiene practices reduces the risk of product contamination.
5. Properly cleaned and sanitized equipment will minimize the risk of product contamination during packaging.

to the question, is the CCP under control? We also have a very high level of confidence in our ability to control and prevent the potential hazards.

Table 8-3 lists examples of CCPs where we can achieve only partial control. In the first example, bone, metal or stone collectors which are properly installed and maintained can remove hard foreign objects from the processing stream of a wide variety of foods. While these devices can reduce the risk of hard foreign objects, none are 100% effective in many foods. In the second example, peanuts can be mechanically sorted to minimize the presence of mycotoxin in peanut products. Experience has shown that peanuts can be sorted to consistently meet the federal action level for mycotoxin content, but the sorting process will not prevent the presence of low detectable levels of mycotoxins. In the third example, it is known that, when each step in the slaughtering process for beef, pork, lamb, and poultry is correctly performed by a skilled operator, then the risk of contamination with enteric pathogens can be minimized. The slaughtering process, however, is not conducted under aseptic conditions and there is no kill step to eliminate contamination brought into the processing environment with the animal. Our ability to prevent the risk of contamination is limited. Also, the assessment of whether contamination is being minimized depends upon the knowledge and judgment of the person who is monitoring this CCP.

The fourth and fifth examples relate to one CCP (i.e. preventing recontamination of cooked foods with pathogens). Depending upon the circumstances, such as the food, the processing procedure, and the conditions of production, our ability to prevent recontamination with pathogens can vary from a low level of confidence to a relatively high level of confidence. A number of factors must be controlled to prevent recontamination, but only two will be considered here. First, we know from experience that training and reminding employees about proper hygiene practices will reduce the risk of product contamination. We also know from experience that humans will occasionally forget or ignore the rules. It is human nature to take the short cut whenever it is possible or advantageous. In addition, a turnover of employees will bring in new people with very little experience and, initially, a minimum of training. A second factor which can affect the risk of recontamination is how well the processing equipment and environment are cleaned and sanitized. Some producers have learned that operating in a dry environment can further reduce, although not eliminate, the risk of listeria contamination of cooked products. Other producers have elected to use an opposite approach and maintain a wet environment with frequent applications of sanitizer. Whether it is for listeria control or another pathogen, the questions arise, how clean is clean, how well is the sanitizing procedure performed, and how dry is dry? These are difficult assessments to make during the short time available for monitoring. Among the other factors which must be controlled to minimize the risk of recontamination are equipment design, plant layout, product flow, and personnel flow. In each of the foregoing examples of CCPs which provide partial control, the assessment of whether the established criteria are met depends largely upon judgment and experience. It is only through some activity of verification (e.g., equipment swabs, product testing) that data can be developed to verify whether the original assessment made during monitoring was correct. Thus, our ability to assess control at a CCP is not always as good as we would like. Also, there are some CCPs where we can minimize but not prevent a hazard.

Adjusting process to maintain control

Let us return now to the first activity of corrective action; namely, to adjust the process to maintain control. The factors which are monitored are often the same factors which must be adjusted to maintain control. These include time, flow rate, temperature, humidity, pressure, vacuum, etc. (Table 8-4). Examples of adjustments which can be made in a process to maintain control are listed in Table 8-5. The first activity of correction action, then, is to use the data from monitoring to anticipate and prevent problems. Statistical process control is a very useful management tool to achieve this objective. A brief discussion of the use of statistical process control in a HACCP Plan is available (ICMSF 1988).

The primary objective of HACCP is to prevent problems. HACCP is a system in which potential hazards are identified and then strategies are developed to

TABLE 8-4   Examples of Factors Which
are Commonly Adjusted to
Maintain Control at a CCP

Time, Flow rate	Chlorine content
Temperature	pH, Acidity
Humidity	Personnel practices
Pressure	Ingredient concentration
Vacuum	

TABLE 8-5   Examples of Adjustments in a Process to Maintain Control

1. The temperature of products flowing through the holding tube of a pasteurizing system is automatically monitored and adjusted. If control is lost, the flow diversion valve is activated.
2. Continuing to cook an oven load of product to achieve the required minimum internal temperature.
3. Holding a vat of milk until the required titratable acidity is reached before increasing the temperature in cheddar cheese production.
4. Monitoring the time and temperature for holding frozen pork to assure the destruction of trichinae.

prevent their occurrence. If problems continue to occur, then the HACCP Plan was either insufficiently developed or the Plan is not being followed as designed. When properly implemented, HACCP is a means to reduce hazards by preventing problems. The concept of using HACCP to prevent problems should be further expanded in future revisions of the NACMCF's HACCP document (NACMCF 1990).

While it is important to know when a deviation occurs, the primary objective of HACCP is to prevent deviations. The importance of this aspect of corrective action is considered in ICMSF Book 4 (ICMSF 1988) where it is stated, "Monitoring must be able to detect any deviation from specification (loss of control) and to provide this information in time for corrective action to be taken to regain control of the process before there is need to reject the product" (ICMSF 1988). *Ideally,* if the process is properly monitored and controlled, there would be no need for Principle No. 5 as described in the USDA document. The ultimate goal of a HACCP Plan is to achieve zero defect. Even in CCPs where hazards can be minimized but not prevented, the attitude of those involved in the operation should be to strive toward zero defect. However, in food processing operations there will be situations where deviations occur from unexpected situations affecting equipment, ingredients, processes, etc.

It is important that some*one* is assigned the responsibility to adjust the process and inform others if a deviation occurs. The word, *one,* must be emphasized. If this responsibility is not clearly assigned to one individual, then it is very likely that the information obtained from monitoring will not be used to adjust the process. One individual must be responsible.

An example of how the corrective action step fits into a HACCP Plan appears in Table 8-6. The example involves the production of a turkey roll. The step in the process is cooking. This step is a CCP as indicated in the box below the column headed, CCP. The product must be cooked to an internal temperature of 160°F or higher. The oven operator is responsible for monitoring each oven load. In this example, corrective action consists of continuing to cook the product until the minimum of 160°F is achieved. This is the responsibility of the oven operator.

TABLE 8-6   Example of Corrective Action in a HACCP Plan (Product: Turkey Roll)

Procedure	CCP	Criteria	Monitoring: Responsibility Frequency	Corrective Action	Verifying: Responsibility Frequency	Records
Cooking	CCP	Cook to ≥160°F internal temperature	Oven operator, Each oven load	Continued cooking until 160°F	Supervisor, Process control, QA; Daily	Cook chart initialed by operator. Save for 1 year

There are some who believe that everything which can possibly go wrong at each CCP should be identified and then an action plan should be specified for each possible problem. This approach will create a very detailed document, but one which will not likely be used. It is very difficult to anticipate all the possible problems. It is better to simply identify the most likely problems and then encourage free communication if something out of the ordinary occurs.

The turkey roll example raises an interesting point. An internal temperature of 160°F is required to satisfy the USDA regulations for the production of a noncured poultry product. This far exceeds the temperature required for safety. It should be possible to produce microbiologically safe poultry rolls using an internal temperature of 145°F or lower as is required for roast beef. The 160°F requirement has its origin in research reported by Wilkinson et al. (1965). It was concluded from the research that 160°F would be required to assure the destruction of enterococci which, at the time, were believed to cause foodborne illness. Since enterococci are no longer considered to be foodborne pathogens, it should be possible to alter the requirement of 160°F. In terms of HACCP, this raises an issue which must be resolved by each company in its own fashion. The current trend in the U.S. is to limit critical control points to assure food safety. One approach might be to conclude that 145°F is required for the microbiological safety of a poultry roll. Thus, 145°F would be listed in the HACCP Plan as the criterion for safety. The 160°F requirement is not a CCP for safety but a regulatory requirement.

Noncompliance product and correcting cause of deviation

The second and third activities of corrective action occur when control is lost and a deviation occurs. Deviation can be defined as "when a product or process fails to meet established criteria." Since the definition for CCP in the NACMCF

(1990) document is limited to safety concerns, a deviation at a CCP is a safety issue and may become a regulatory issue depending upon the circumstances. Seven possible corrective actions can be taken when control is lost. (1) If necessary, stop the operation. (2) Place all suspect product on hold. (3) Provide a short-term fix so that production can be safely resumed and additional deviations will not occur. (4) Identify and correct the root cause for failure so that future deviations will not occur. (5) Deal with the suspect product. (6) Record what happened and the actions taken. (7) If necessary, review and improve the HACCP Plan.

There are only a few options for the disposition of suspect product. The first option is to release the product. While this is an option, it is not the wisest approach if there is a question of safety involved. The second option is to test the product to verify whether the product is safe to release. This is a fairly common approach to assess the acceptability of suspect product. Several sampling schemes have been proposed but will not be discussed in this text. Sampling suspect product is a topic which cannot be treated lightly. This option must be approached with caution because the statistics do not favor the detection of a defect which is present at a very low incidence in a product. The third option is the divert the product to a safe use. For example, pasta, eggs, or cooked chicken contaminated with salmonellae should not be used by a processor who will use them as ingredients in a salad product. On the other hand, it would be acceptable to use these materials as ingredients in the manufacture of canned, retorted, shelf-stable products or other products which receive a kill step which is adequate for salmonellae destruction. The fourth option is to reprocess the product. The fifth option is to burn, bury, or otherwise destroy the product.

Reaching a decision on the appropriate disposition of noncompliance product can be influenced by several factors. First is the severity or the seriousness of the hazard. For example, does the hazard involve spoilage or botulism. The second factor is risk. This is the likelihood of the hazard occurring. Is it one chance in a million or is it likely to occur every time the deviation occurs? The third factor is how the food will be stored, shipped, and prepared. The fourth factor is who will prepare the food. The fifth factor is who will consume the food. Each of these factors and, perhaps, others should be considered before reaching a recommendation on the disposition of the product.

The concept of risk and its relationship to the corrective action which should occur when control is lost has been adopted by the Food and Drug Administration in its frozen dessert processing guidelines (FDA 1989). Although the severity of the hazard has not been considered in the general classifications, the level of risk has been divided into three categories: low, moderate, and high risk. The recommended corrective actions depend upon the level of risk. A summary of the FDA guidelines appears in Table 8-7.

Who should recommend the disposition of suspect product? There are two

TABLE 8-7   Corrective Action Recommended by FDA for the Production of Frozen Dessert[a]

Level of Risk	Risk Assessment and Recommended Corrective Action
High risk:	(a) There is a high level of risk that a hazard will occur which directly impacts product safety. A high level of control is needed to assure that these problems do not occur. (b) *Action Priority*—No product should be processed until the problem is corrected. If appropriate, product should be placed on hold and tested. If the product fails, then appropriate action is required.
Moderate risk:	(a) There is a moderate level of risk that a hazard will occur. The potential hazard can occur if other factors (e.g., temperature abuse, failure to meet certain criteria) also are not met. Timely monitoring is required because these problems could result in a risk to product safety. (b) *Action Priority*—Product can be produced, but the problem should be corrected within a short period of time (e.g., a few days or weeks). Specific additional monitoring is needed until the correction has been accomplished.
Low risk:	(a) There is a low level of risk that a hazard will occur. Significant risk of a hazard would result only after extensive abuse or other extenuating circumstances. Monitoring is needed only on an inspection or random-checking basis. (b) *Action Priority*—Product continues to be produced. These problems should be corrected, for example, when production schedules permit. Routine checks should be made to assure the status has not changed to moderate or high risk.

[a]Adapted from the FDA Frozen Dessert Processing Guidelines (FDA 1989)

types of experts involved in a HACCP Plan. The first is an expert in HACCP. The characteristics of an expert in HACCP are listed in Table 8-8. These are individuals who understand the principles of HACCP, can serve as a facilitator during the development of a HACCP Plan, and can manage, review, or verify a HACCP Plan. These individuals, while they are competent and have a working knowledge of the HACCP Plan, may not be qualified to recommend the disposition of suspect product. The second expert involved in HACCP is an expert in hazard analysis. The qualifications of an expert in hazard analysis appear in Table 8-9. This individual can be a chemist, a toxicologist, a microbiologist, a

TABLE 8-8   Expert in HACCP

- Understands the principles of HACCP
- Can serve as a facilitator during the development of a HACCP plan
- Can manage, review, or verify a HACCP plan

TABLE 8-9   Expert in hazard analysis

1. A specialist (e.g, chemist, toxicologist, microbiologist) in the hazards of concern.
2. Has the knowledge and experience to correctly:
   (a) Identify potential hazards
   (b) Assign levels of severity and risk
   (c) Recommend controls, criteria, and procedures for monitoring and verification
   (d) Recommend the disposition of product when criteria are not met
   (e) Recommend research related to a HACCP plan
   (f) Predict the success of a HACCP plan

veterinarian, or some other specialist who is knowledgeable in the particular hazard of concern. These individuals have the knowledge and experience to *correctly* identify the hazards, etc., as outlined in Table 8-9. The recommendations of the expert in hazard analysis can significantly influence the cost to implement a HACCP Plan and the cost of the operation involved. A very conservative individual or one who is not truly knowledgeable in the product or process in question could impose unduly strict, costly requirements. A "process authority" has been proposed as an expert in hazard analysis. The qualifications of a process authority have not been defined and it is uncertain whether this individual is capable of correctly recommending the disposition of suspect product. Relative to corrective action, it is of interest that five of the six activities of the expert in hazard analysis in Table 8-9 are involved with the prevention of problems. Only one activity is concerned with the disposition of product produced when control is lost and criteria are not met.

Once a recommendation is developed for the disposition of suspect product, someone must make the final decision of what will be done with the product. This decision is normally made by the individual who has profit/loss responsibility for the product in question or someone in a higher position within the company. It has been my experience that these individuals have always accepted the recommendation from the expert in the hazard in question. This decision may be communicated to the appropriate regulatory agency for the food in question. Under certain circumstances, the final decision might be jointly arrived at by the company and the agency.

References

U.S. Food and Drug Administration (FDA). 1989. *Frozen Dessert Processing Guidelines.* 1st edition. Milk Safety Branch, Division of Cooperative Programs, FDA, Washington, DC.
International Commission on Microbiological Specifications for Foods (ICMSF). 1988. *Microorganisms in Foods. 4. Application of the Hazard Analysis Critical Control Point*

(HACCP) System to Ensure Microbiological Safety and Quality. Blackwell Scientific Publications, London.

National Advisory Committee on Microbiological Criteria for Foods (NACMCF). 1990. *HACCP Principles for Food Production*. USDA-FSIS Information Office, Washington, DC.

Tompkin, R.B. 1990. The use of HACCP in the production of meat and poultry products. J. Food Prot. 53: 795–803.

Wilkinson, R.J., Mallmann, W.L., Dawson, L.W., Irmiter, T.F., and Davidson, J.A. 1965. Effective heat processing for the destruction of pathogenic bacteria in turkey rolls. Poultry Sci. 44: 131.

9

Effective Recordkeeping System for Documenting the HACCP Plan

K. E. Stevenson and Bonnie J. Humm

Principle 6. *Establish effective recordkeeping systems that document the HACCP plan.*

INTRODUCTION

Records are written evidence through which an act is documented. The act of keeping records assures that this written evidence is available for review and is maintained for the required length of time.

Since part of the HACCP plan includes documentation relating to *all* critical control points (CCPs) identified in a food establishment operation, records are an integral part of a working HACCP system. All physical or chemical measurements of a CCP, any action on critical deviations and final disposition of any product must be correctly documented and kept on file.

All records that relate directly to these CCPs are to be made available to government inspectors upon request, for the HACCP plan clearly delineates which records fall into this category. Records that deal with the functionality of the HACCP system and other proprietary information are not necessarily required for review by these regulatory agencies. Records are the only reference available to trace the history of an ingredient, an in-process or a finished product. If questions arise concerning the product, a review of the records may be the only way to ascertain or even to prove that the product was prepared and handled in a safe manner in accordance with all the HACCP principles outlined in the company's HACCP plan.

In addition, recordkeeping is a tool or mechanism by which an operator may learn of an equipment malfunction that could violate a critical factor and allow that operator to correct a potential problem. A record of this type provides a

83

dual function by providing a history of the machine's performance as well as an action taken to correct a deviation.

Record reviews must be conducted in-house by qualified staff members as well as by outside HACCP authorities such as consultants or regulatory inspectors in order to assure strict compliance with the criteria set at the CCPs. Careful review of well-documented and well-maintained records is an invaluable tool in indicating possible problems and allowing corrective action to be taken before a product health risk occurs.

Regulations specify that copies of all required records be retained at the processing facility for one year from the date of manufacture and for two additional years at a reasonably accessible location. Certainly it is prudent to keep all records at least as long as the intended shelf life of the product should this time extend beyond the three year requirement.

Reasons for keeping records

The reasons for keeping HACCP records relate to evidence of product safety with regard to the present procedures and processes, assurance of regulatory compliance, and ease of product traceability and record review.

Well maintained records provide the best evidence that procedures and processes are being followed in strict accordance with HACCP requirements. Adherence to the specific critical limits set at each critical control point is the best assurance of product safety. Documenting the data of those measurements results in permanent records regarding the safety of those products.

During regulatory compliance audits, company records may be the single most important source for data review, and, depending on the thoroughness of the records, inspectors can easily ascertain the adequacy of processes and procedures used at the facility in question. More importantly, accurate records also provide plant personnel with this documentation of compliance.

Since HACCP records focus only on safety-related issues, problem areas can be quickly identified because these records provide an uncluttered view of product safety issues. All HACCP records should be kept separate from quality assurance documents so that regulatory compliance officer will view only the product safety records during HACCP audits. In addition, HACCP records assist in identifying lots of ingredients, packaging materials and finished product should a product safety problem occur requiring a market withdrawal.

TYPES OF HACCP RECORDS

Critical control point (CCP) records

These records document the identification of specific hazards and the accompanying risk assessments associated with each CCP. These hazards could be found in an ingredient, a packaging component or the process and may be of a biological, chemical or physical nature.

A diagram or flow chart of the entire manufacturing process with each hazard identified and a critical control point identified would be an example of this type of record. Since each CCP requires a risk assessment, proper documentation regarding the thought process in determining the degree of risk associated with each hazard should be included in this category.

Once the CCPs have been assigned, a decision regarding the degree of control attainable at each CCP must be made. The criteria behind this decision should then be fully documented and made part of the HACCP records.

Records associated with establishing critical limits

In order to support the critical limits established for each CCP, studies may have to be conducted and experimental data collected. The rationale used to support the conclusions are important and should be included in these supporting data. Any pertinent literature regarding the history of such criteria enhancing product safety should also be included in this type of record. The precision and accuracy of all test methods used in the measurement of critical limits must be well documented before making such tests part of the supporting documents for the HACCP program.

There are always normal and/or acceptable fluctuations in the data collected from most operations and these fluctuations will be apparent on the records. It is imperative that the individual responsible for recording the critical control point data knows the difference between normal fluctuations and an indication of loss of control at any critical control point location. These guidelines must be clearly stated and the limits printed on each CCP record or data sheet for easy reference by the operator or attendant.

Examples of automated equipment records include circular charts showing control of process temperature and strip charts from measurement panels attached to diversion control valves for assuring proper sauce fill temperatures. Spot checks of continuous inspection activities would include raw ingredient inspection, sanitation swab sampling and pre-operations sanitation checks. Proper documentation of continuous monitoring systems will result in various charts, check lists and laboratory analysis sheets.

Discontinuous inspection known as attribute sampling may be used in tests for chemical or physical parameters. This type of inspection is based on statistics, and it requires accurate documentation forms for each lot sampled.

Records associated with deviations

The failure to meet a required critical limit for a critical control point is termed a deviation. All deviation procedures must be documented in the HACCP plan and agreed to by the appropriate regulatory agency prior to approval of the plan. Each deviation requires a corrective action which must eliminate the actual or potential hazard and assure the safe disposition of the product involved. This requires a written record identifying the deviant lots and the holding of that product pending completion of the appropriate corrective actions. A Hold Summary could be the master form for these deviations with supporting documentation kept on separate file for a reasonable period after the expiration date of the product.

The final disposition and handling of all process or product deviations should be very detailed including an accurate accounting of all units. This includes product destroyed as well as product reworked or returned to stock.

Since HACCP deviations are of a product safety nature rather than quality, these records should be kept in a separate file apart from quality assurance or regulatory requirement records. This facilitates record review by staff as well as regulatory personnel.

Records and verification

Verification is an integral part of the total HACCP system because it provides feedback through internal and external auditing of existing data. Both the producer and the regulatory agencies have vital roles in verifying HACCP plan compliance. In addition to visual inspection of the operations, verification may include review of all HACCP records for compliance with the HACCP plan. These external inspections may be routine or they may be the result of a consumer complaint if food safety is at issue. Keeping records in an orderly fashion and easily retrievable will make in-house review an easy task and also facilitate external inspections.

Spot checks or sample analyses are also an established verification procedure which may require specific documentation from the processor regarding lot or code numbers. These checks may serve to validate an existing ingredient warranty or guarantee kept on file and serve to substantiate the assignment of certain critical control points in the HACCP system. In addition, the result of such analyses provides evidence of the adequacy of present procedures used in-house for these critical factors.

Since proper calibration of existing equipment is (paramount) critical to the accuracy and precision of any analyses, record review may alert plant personnel of an existing or potential problem so corrective action can be taken.

In-house verification procedures require a comprehensive record review on a routine basis in order to assure full compliance with the HACCP program. All HACCP records must contain the following information:

Title and date of the record
Product identification (code, including time and date)

Materials and equipment used
Operations performed
Critical criteria and limits
Corrective action to be taken and by whom
Operator identification
Data presented in an orderly format
A place for the supervisor's initials

RECORD REVIEW AND RETENTION

Records dealing with critical inspection points must be reviewed on a daily basis by a designated responsible individual. All records should be initialed and dated as they are reviewed. This review must then be followed up by a review of any deviations or irregularities. Any deviation from standard documentation procedures must be brought to the attention of the individuals responsible for filling out the reports and these deviations should be immediately corrected.

Any anomalies must be thoroughly investigated for potential problems or trends, and in this regard record review becomes a preventive measure for assuring product safety. When this review reveals or identifies an inadequacy in the recordkeeping or normal monitoring procedures, then existing parameters must be reviewed and updated. Because implementation of the HACCP program results in a dynamic system, continual updating and improvement are then necessary and well warranted. Additional critical control points not addressed in the original program may become evident as the record review and verification process is applied to each component of the system.

THE HACCP PLAN

The HACCP Plan is a written document which delineates the formal procedures to be followed in accordance with the seven principles. It may consist of a HACCP manual or working document, appropriate HACCP test methods or SOPs, and a Master file containing all background documentation and HACCP records.

The manual should include all the elements of the HACCP plan, flow charts, procedures, test methods, documentation requirements and a copy of all required data sheets including instructions on how to fill them out.

Plant personnel such as line workers or laboratory analysts should have copies of all SOPs or appropriate test methods for which they are responsible. This will enable them to properly execute their individual HACCP assignments.

Any changes to the HACCP plan must be immediately reflected in the HACCP manual. All charts should have issue numbers so critical limits and instructions are kept current. Outdated sections and documentation forms should be immediately discarded to avoid confusion. A periodic review of departmental HACCP forms and procedures may be necessary to assure compliance with the primary HACCP plan.

When revisions are made and sent to their respective departments, it is advisable to have routing slips attached so that the individuals responsible for the implementation of those revisions are properly notified. There is nothing worse than having outdated HACCP documents still on file and in use at a facility purporting to be operating under a HACCP system.

Investigative reviews should be conducted by staff personnel prior to regulatory inspection in order to identify weaknesses in the documentation or recordkeeping system. Having a well organized system for documentation will show that a company is in control of the operation in general and is in control of product safety issues specifically. It is important that all in-house record reviews be well documented with all deficiencies noted and corrective action clearly outlined. When problems continue to occur in certain areas, there is a written record of the reasons and the proposed solutions.

Retention of records

Regulatory requirements for retention of records varies among regulatory agencies and locales. HACCP records should be held for at least a year, while any records required by law to be kept longer than one year should be kept the legally mandated period. The shelf-life of a product also needs to be taken into account in establishing retention guidelines. The regulations specify that copies of all required thermal processing records, records of pH measurements, process deviations and other "critical factors" shall be retained at the processing facility for one year and two additional years in an accessible location. The HACCP records dealing with critical factors (CCPs) definitely fall into this category.

Regulatory access

The types of records utilized in the total HACCP system include records on ingredients and packaging materials, processes and controls, packaging requirements, storage and distribution. Records that deal with the management or function of the system itself and proprietary information would not normally be made available to the regulatory agencies. Records that clearly relate to product safety are already identified in the HACCP program and are therefore subject to scrutinization by regulatory authorities. Having these records well organized make data retrieval an easy task for both internal and external audits.

Agencies also have legal authority to access non-HACCP records that deal with current laws, regulations and other guidelines. Proving compliance with federal regulations for low-acid foods, such as 9 CFR Part 318 (381) "USDA

Canning Regulations" and 21 CFR Parts 113 "Thermally Processed Low-Acid Foods Packaged in Hermetically Sealed Containers," mandates adequate recordkeeping procedures.

All food industry personnel are responsible for food safety including company executives and managers. Under the Federal Food, Drug and Cosmetic Act criminal penalties can be assessed against all responsible individuals even though those in charge were unaware of the violations and had no intent on violating the Act. To meet these responsibilities, personnel must be aware of both past and present operations and therefore records must be accurate in order to reflect the actual operational conditions.

Personnel responsible for documenting HACCP records should never pre-record data in anticipation of the actual data or postpone making entries and relying on their memory. These records may be the company's only proof that a critical factor was controlled or that corrective action was taken to assure the safety of the product. Any modifications to the existing data should never be erased, but lined out and corrected with the responsible individuals initials alongside the change.

To be used effectively, HACCP records should be on standardized forms for the company and reviewed regularly by a responsible individual for completeness. A thorough review must ensure that all critical factors have been satisfied and are accurately documented.

SUMMARY

Accurate recordkeeping is an integral part of the successful implementation of a HACCP plan. The safety of a company's products depends on such documentation, and customer satisfaction and product liability demand it.

Management, supervisors, on-line workers, and regulatory personnel are all responsible for the safety of our food products, and therefore have a primary role in assuring that all HACCP records are kept accurately and that these records reflect the actual operating conditions. Assuring compliance with existing and newly promulgated regulations requires each processor to keep updated on the regulations governing product safety.

10

Verification of the HACCP Program

Gale Prince

Principle 7. *Establish procedures for verification that HACCP system is working correctly.*

IMPORTANCE

Verification is a very important step in a successful HACCP program. The purpose of the verification step is to confirm through documentation that the HACCP plan is followed as outlined. The HACCP program is designed to concentrate on food safety elements and prevent food safety problems from reaching the consumer. The verification step provides assurance the HACCP program is achieving the established objective of food safety.

Food safety is an invisible challenge. During the food production process you may be confronted with biological, chemical, and physical hazards which cannot be seen with the naked eye. This challenge will not diminish but will increase as we direct our efforts to meet the growing consumer demands, for newer convenient type foods that meet consumer expectations. Food safety will not just happen, it must be built into a product.

PREVENTION

The HACCP program allows us to apply scientific knowledge to a product and a process in order to achieve product safety. It provides the mechanism of control which can be used effectively in managing a food safety program. Implementation of the HACCP program ties together all of the in-plant assets associated with producing a safe product. The HACCP program is a concept of prevention, through the establishment of control measures, based on scientific elements of a product and the manufacturing process. This allows a food processor to anticipate potential food safety problems and to take corrective action, before a situation goes unnoticed and becomes a problem.

TEAM APPROACH

For the HACCP program to be successful it must be a team approach. The plant food safety team must be involved from the development step through the verification step. By involving all the employees in the development of the HACCP program, you expand their knowledge of the concept as well as develop a team committed to food safety.

This communication is very important in developing a food safety team, knowledgeable in the everyday production of a safe food product. These employees must know and understand the importance of the critical control points. If the HACCP program is properly established and followed, the plant's food safety program will be a "preventive" program.

ESSENTIAL ELEMENTS

The verification process is an essential element in keeping a HACCP program functioning properly. When studying food safety failures, employee complacency is frequently a contributing factor. Individuals responsible for food safety become complacent and are content with less that what is acceptable. While product quality can vary and still be acceptable, food safety cannot. The verification step is to ensure that those food safety measures identified in the HACCP program do not vary beyond their established limits.

Each HACCP program developed is unique, in that it is based upon the product produced, product formula, processing equipment used, the packaging material used, and consumer use. Thus one HACCP program does not fit all products or all plants. The concepts discussed in this book can be used to develop a customized HACCP program for any product or food processing operation.

The verification step is made up of five parts:

1. A review of the HACCP plan,
2. Compliance with the established critical control points,
3. Confirmation of compliance with procedures for handling of deviations and records,
4. A visual inspection of the operation while a product is in production,
5. A written report.

To begin, identify a product and its production line. The verification step should involve a team of people. This may include production line supervisors, plant management (quality control manager, plant superintendent, plant engineer, etc.), the Corporate Quality Assurance Department or the use of a consultant experienced in HACCP. The latter may provide a new insight to the HACCP program.

CRITICAL CONTROL POINT VERIFICATION

The verification procedure should be a routine part of the daily production process and also a detailed review of the entire HACCP plan. On a routine basis, verification of the established critical control points must be done daily on each production lot. Also, a more thorough verification process should be performed, on an announced schedule, when there are to be changes in the product or process, when new information becomes available from your own laboratory, when new scientific research information is published, or as information is identified by a regulatory agency. For example, the need for more frequent verification of the HACCP program may come from in-plant laboratory test results, shelf life studies, consumer complaints, published product recalls, reports from Centers for Disease Control investigations, etc. Regulatory inspections can also provide information on the need for verification of a HACCP program.

VERIFICATION VS. MONITORING

Verification is different than monitoring. Monitoring is like quality control in that the monitoring step is going on during the process so adjustments can be made in the process before the product leaves the production line. Verification is like quality assurance in that it is a check on the system to confirm that the established critical control points were properly verified and corrective action, if needed, was properly taken. The verification process involves a review of production records covering the critical control points, by plant management to verify the HACCP program has been followed as it was outlined. This process covers the test results and records for each critical control point.

HACCP PROGRAM VERIFICATION

Periodically, verification of the entire HACCP program must be conducted. The frequency may depend upon the complexity of the product, the degree of risk associated with the product and when process changes are made. For example, baby food or food designed for the aged may require more frequent review. A full HACCP verification review is more than just reviewing the critical control point records. It involves a complete review of the entire HACCP plan. This review may be conducted on an announced or unannounced schedule. The HACCP program verification process involves the review of all elements of the program. This review is essential for determining if the plan is current and provides food safety assurances as written.

- the written HACCP plan,
- critical control point records,

- deviations and corrective action to be taken when a deviation occurs,
- raw material specification compliance,
- processing equipment compliance with the plan,
- verification of the testing equipment/calibration to standard.

Although not all-inclusive, these are some of things which should be evaluated in regard to each of the above mentioned elements.

Raw ingredients

- specifications
- approved supplier
- ingredient sampling
- any changes in raw ingredients and suppliers

Receiving and storage

- temperature controls
- humidity controls
- quarantine program followed
- stock rotation

Processing

- verify the process formula during the process
- review processing steps
- are established control measures being followed
- temperature/chemical control measures
- time/temperature control
- documentation of these control measures
- concentration of cleaners and sanitizers

Processing equipment

- is the flow diagram for the operation current
- is the same equipment being used as when the HACCP plan was established
- any equipment modifications or changes noted
- do the process control mechanisms such as thermo controls, equipment drive belt pulleys, chain sprockets, product piping changes, etc. comply with the HACCP plan
- were systems changes discussed with the food safety team prior to making any changes

Cleaning and sanitizing

- chemical concentrations of cleaners and sanitizers
- clean in place (CIP) control charts
- time/temperature/pressures

Control devices

- is the accuracy of the measuring device being evaluated on a schedule identified in the HACCP plan
- is the verification done against a standard
- is the frequency as outlined
- are seals in place on the critical process control device(s)
- are routine checks made of the seals on the control devices and documented
- do the control charts accurately document the critical control points
- do the control charts accurately document product, code, amount of product produced, who processed the product

Packaging materials

- have packaging materials been changed
- is the production code legible and does it correspond to the production batch and the production records
- are the established shelf-life test results followed
- have finished product handling procedures changed
- have consumer use directions changed

The packaging specifications should be compared to the packaged product off the line with the consumer view in mind. This should include product identity, ingredient statement, product code, and consumer use directions to verify that no changes have occurred from when the HACCP program was established.

When conducting HACCP program verifications one should not overlook the production employees point of view as they are an excellent source of information on the HACCP program. This is an opportunity not only to gain insight into the process but may also be used to determine employee knowledge level of the product and the HACCP system. This may determine HACCP training needs. The above verification elements are not all inclusive for every product, process or plant but provides a guideline to the elements of conducting a verification review of the established HACCP plan.

FINISHED PRODUCT TESTING

When discussing HACCP program verification, there are different philosophies on the subject of sampling finished product. If the HACCP program has been properly set up and followed there should be minimal need for finished product testing. Food safety cannot be tested into a product by laboratory testing of finished products. Food safety must be designed and built into product through the HACCP concept. The tests done during the monitoring step should be quick to clear the product before the product leaves your control. Verification testing may be done after the fact, and may depend upon the product. If it is a shelf stable product designed especially for infants or immuno-compromized and is a high risk product, your program may require verification testing before the product is shipped. For example, the FDA low-acid canned food regulations also requires an incubation test of finished product to verify the safety of the lot.

RECORDS

An effective HACCP program depends upon records that document the safety of the process. Inspections of production records have frequently shown voids or incomplete documentation of a process. The verification step must review all critical control point records and the HACCP plan. The records must properly identify the product, product code, amount of product produced, temperature of the process or chemical concentration, etc. Proper documentation of records for critical control points must be confirmed by a member of management.

Documentation of process deviations are an important part of a HACCP program. Corrective action must be documented at the time corrective action was taken. Product impoundments should be listed and also cleared when the corrective actions were taken. Handling of rework material should also be covered in the verification step. Record retention is another subject of a HACCP verification step. The process records must be maintained to verify the safety of the product to cover the entire shelf-life of the product plus a reasonable period to allow for consumer consumption. The HACCP plan needs to be a permanent part of the product file. A record retention program must be outlined and followed. A review of the record retention program needs to be a part of the verification step.

WRITTEN REPORT

Each time the verification step has been completed, a written report must be prepared *to certify the HACCP plan is being followed as outlined*. Deviations from the HACCP plan must be listed in the report and discussed with plant management. A follow up report must document the corrective actions taken in response to each of the deviations identified. The follow up steps should be completed in a timely manner. Again, this is an official record of compliance with the HACCP program documenting food safety measures and is a critical point in the verification step.

The implementation of a HACCP program builds employee knowledge of your product and operation. The verification step continues to build that knowledge level each time it is conducted for everyone involved.

The HACCP program is a management tool to ensure a safe product. The verification step provides confirmation that the HACCP program is effective and working properly in meeting consumer expectations of a safe food supply.

11

Control Points and Critical Control Points

John Humber

INTRODUCTION

Control points are an integral part of a food processor's comprehensive product control system and can be used, together with HACCP, to help ensure that the consumer receives a safe food product with consistently good quality. The integration of control points with HACCP to maintain the safety and quality of foods has been described by Sperber (1991).

The National Advisory Committee on Microbiological Criteria for Foods (NACMCF) defines a control point as any point in a specific food system where loss of control does not lead to an unacceptable health risk (NACMCF 1990). But as the food scientist begins to develop a HACCP system to control the biological, chemical and/or physical hazards of a food processing operation, difficulty may arise when the time comes to distinguish between control points and critical control points. As a result, the food manufacturer very often has more critical control points than expected, and perhaps many more than needed. This chapter will present a method of more easily identifying control points and understanding their role in addressing the safety, quality and regulatory aspects of food production.

DEFINITIONS

As part of the process of more easily identifying control points, and then separating control points from critical control points, it is necessary to slightly modify the currently accepted definitions (NACMCF 1990). This revised definition would read: Critical Control Point—Any point in a specific food system where loss of control may result in a *high* probability of a health risk. The revised definition of critical control point now contains the concept that if control of the point is lost, there must be a *high* probability that a health risk will occur. On the other hand, if control is lost but the risk is *low* that a health risk will occur, then the concern should be classified as a control point. A Control Point definition would then be: Any point in a specific food system where loss of control may result in an economic or quality defect, or the *low* probability of a health risk occurring.

COMPREHENSIVE PRODUCT
CONTROL SYSTEM

If the definitions of critical control point and control point are then applied to the development of a comprehensive product control system, where all phases of safety, quality and regulatory are addressed, then Fig. 11-1 could be used to illustrate the overall relationship between the high and low risk concerns for a specific food product. Listed under HACCP are the critical control points that if not kept under continuous control will *very* likely lead to a health risk. Thus, the risk should be high for those points classified as critical control points.

The United States Department of Agriculture has appropriately placed a high priority on food safety, and proposes to use HACCP to control only the safety concerns of a food process, rather than as a system to manage quality and economic issues as well (USDA 1990a, b). This allows the food processor, in turn, to focus on the control of key safety issues for a given product and process. The National Food Processors Association supports the concept that HACCP

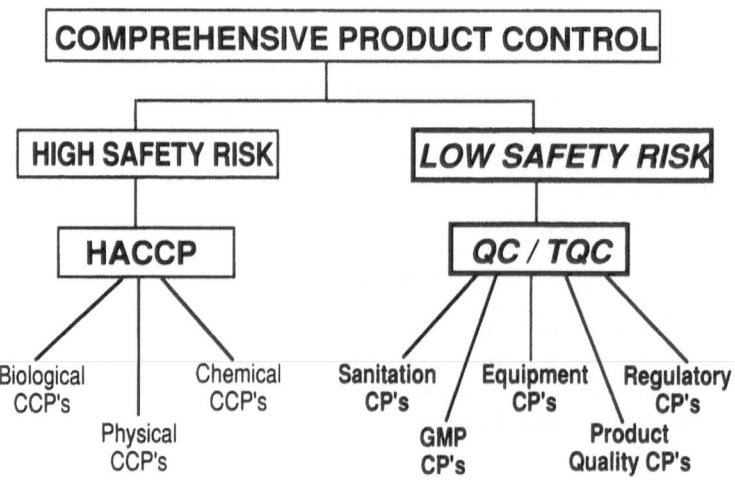

FIGURE 11-1.   Relationship between high risk and low risk concerns for a food product.

should be limited to those critical control points that are necessary to prevent illness (NFPA 1989). HACCP systems stressing food safety have been developed for chilled foods (CFA 1990; Bryan 1990), refrigerated foods (NFPA 1989) and dairy products (ABI/NCI 1990). A HACCP plan for use by the consumer to help prevent foodborne illness has also been published (USDA 1989).

In Figure 11-1, the points in a food production operation that have a low safety risk, and therefore are not controlled by HACCP, are classified as control points. The probability that these points would cause a health risk, if not continuously controlled, is low. Control points in a food production operation are usually placed under a total quality control program. Examples of areas where control points are often found include the sanitation of production lines, production plant good manufacturing practice procedures, equipment maintenance, product quality attributes, and certain areas that are controlled by federal, state or local regulations. It is important to understand that certain lower risk safety concerns can be placed under the quality control heading, separate from HACCP. HACCP continues to encompass and control the major safety issues, either biological, chemical or physical, but only those of high risk and that are very likely to cause a problem if control is temporarily lost. Low risk, or low probability safety concerns can, and should be placed with quality and economic control points, separate from HACCP.

Decision tree

The decision tree presented in Fig. 11-2 suggests a way to identify control points and then to separate them from critical control points. It asks the question: "If I lose control, is it *likely* that a health risk will occur". If the answer is yes, then the point of concern in the process should be classified as a critical control point. On the other hand, if there is a point in the production flow that is somewhat

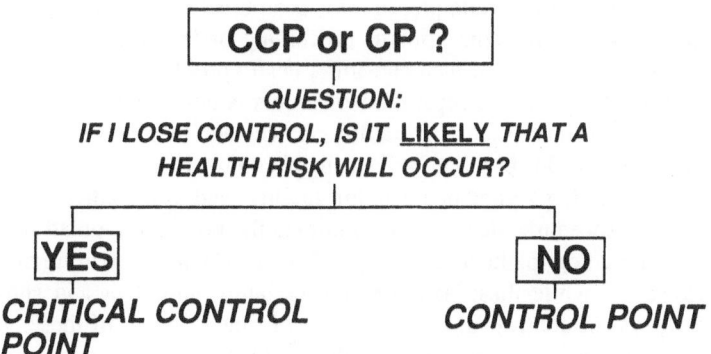

FIGURE 11-2.   CCP, CP decision tree

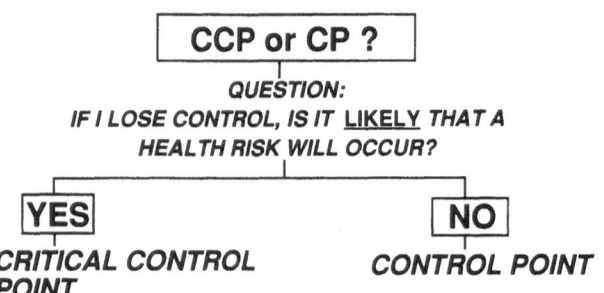

FIGURE 11-3.

related to safety, but the probability or likelihood is low that someone will become ill or injured if control is temporarily lost, then that point should be placed in the comprehensive product control system as a control point, but separate from HACCP.

The eight examples listed in Fig. 11-3 illustrate the separation of low safety risk control points from higher risk critical control points achieved by answering the suggested question. It is obvious that if the two thermal processing critical control points were not continuously controlled, the chance of having a health risk occurring would be high. Correspondingly, a footbath in a canning facility and a handwash station in a frozen vegetable plant certainly relate to safety and can help to minimize contamination of raw ingredients with pathogens, but a temporary loss of control would present only a minimal risk. Although pathogen growth in the raw ingredients prior to processing or freezing is possible, the likelihood of it occurring and then causing a health problem is really very low.

As mentioned earlier, the separation of points is not based on the presence or absence of a safety issue, but rather on whether there is a high or low probability that a health problem will occur if control is temporarily lost. The chlorination of mixing kettles in a canning facility, and pest control in a cheese plant are both procedures designed to minimize the growth or spread of micro-organisms, including pathogens, in the food production environment (Fig. 11-3). However, while these activities are certainly important and should be maintained, loss of control for a short period of time of either of these controls would not likely lead to a consumer health problem. They are therefore listed as control points.

Two other examples listed as critical control points in Fig. 11-3, refrigeration and metal detection, represent points that if not continuously controlled would likely lead to a health risk for these particular products.

Classification of control points and critical control points

It is important to remember that the classification of control points and critical control points is both product and process specific. For example, certain products or processes may require that metal detection be listed as a critical control point. For another product or process, however, metal detection may be classified as a control point. The difference in classification is based on a risk assessment of the product or process, and the answer to the question asked earlier, "Is a health risk likely to occur if control is lost?"

Control points cannot only help to control low risk safety hazards, but they can also help to control the quality characteristics of a food, such as color and flavor of cheese. Additionally, control points can have a regulatory function, as in the control of the net weight of a box of cereal. As a product flow diagram is assessed, there are several sources of information, or indications, that are available (Table 11-1) or that can be reviewed, to help determine if it is likely that a health risk will occur when the control of a given point is temporarily lost. Various risk assessment methods and analyses of hazards associated with food production have been described previously (Corlett and Stier 1991; Bryan 1990; ICMSF 1988; NAS 1985).

A thorough review of the seven sources of information provided (Table 11-1) will enable a food processor to (1) more easily separate control points from critical control points and (2) identify points in quality and regulatory areas that need to be controlled.

Product history refers to any prior documentation or instance where the product category has been involved in a biological, chemical or physical problem, perhaps resulting in a product recall for public health reasons. A review of this information

TABLE 11-1   Is a Health Risk *Likely* to Occur?

Indications:
   —Product History
   —Consumer Complaints
   —CDC Morbidity and Mortality Weekly Reports
   —1-800 # Calls
   —Process Authority Recommendations
   —Predictive Models
   —Scientific Literature Articles

could guide the food processor in a direction which could help to reduce or eliminate problem, by assigning critical control points to these areas.

If a food processor is beginning to implement a comprehensive quality control plant for an existing product, then consumer complaints or letters from customers could alert the manufacturer to problems associated with the product. For example, a series of complaints of unusual product flavors could lead to a strengthening of the control points that relate to flavor.

Health problems related to a particular product category can be monitored by various means, including a review of the Centers for Disease Control document entitled *Morbidity and Mortality Weekly Report*. This publication provides accounts of certain foodborne illness outbreaks which could alert the processor to potential safety problems for a specific food category.

Food processors often place 1-800 telephone numbers on package containers with the request that consumers call the number if there are concerns related to the quality of a product. An examination of these telephone comments could reveal negative customer feelings that might be easily addressed through a comprehensive product control program and the prudent use of control points.

Perhaps one of the most productive sources of information for food processors is the recognized process authority. Process authorities are typically, but not always, trade associations. Examples of trade associations that could provide recommendations on the safe production of specific categories of food include the National Food Processors Association, the Chilled Foods Association, and the Grocery Manufacturers of America, Inc. Trade associations such as these can be particularly useful in helping to identify regulatory issues that need to be addressed under a control point system.

Predictive models can be useful in assessing the safety risks of new product formulations even before the formulations have been manufactured for the first time in a pilot plant, or sold in a test market. Models can be used to predict both the microbiological safety and quality of a formula, thus giving the manufacturer the opportunity to review the health risks and spoilage potential of a product long before it is eaten by the consumer. Control points relating to safety, quality and regulatory issues can often be more easily identified in advance through the use of predictive models.

Control point lists

As the steps in a process flow diagram are reviewed, and control points are identified and separated from critical control points using the format described in Fig. 11-2, it is often appropriate to develop a complete list of control points for each product.

A list of control points for several typical product situations is presented in Table 11-2. The three control points addressing low risk safety concerns are

TABLE 11-2   Control Points

Low Risk Safety	Quality	Regulatory
Raw milk storage temperature	% Salt in a bread formulation	Ingredient labeling on a cracker box 21 CFR Part 101
Sanitation of a production line	Spice levels in Bar-B-Q Sauce	Addition of Vitamin D to milk 21 CFR Part 131
Air quality in a Supermarket Deli	% Color in fruit drinks	Milk fat requirements for ice cream 21 CFR Part 135

storage temperature, general sanitation and air quality. All three examples represent controls that are important to maintain and monitor, but that are not very likely to result in a consumer health problem if control is temporarily lost. However, in specific product situations, certain individual sanitation procedures may be needed to control high probability safety risks. In these instances, such procedures should be placed under HACCP as critical control points.

The control points for % salt, spice level and % color are examples of points in a system that are necessary to maintain the quality of food products, and to help make products competitive in the marketplace. Consumer correspondence and 1-800 # telephone calls from customers can be useful sources of information to help confirm that control points for quality are in place and are being properly monitored.

The regulatory control points listed in Table 11-2 are typical of those found in a comprehensive product control system. Regulatory requirements may originate from local, state or federal agencies and often address nutritional, economic misrepresentation or adulteration issues. The regulatory control points in Table 11-2 are examples of federal requirements for several different food categories as specified in the *Code of Federal Regulations* (NARA 1990). As with other control points, the examples listed are unrelated to the high probability safety issues which would be controlled within the HACCP system for a given food product.

SUMMARY

In summary, control points in a comprehensive product control system address issues pertaining to low probability safety risks, product quality, and regulatory requirements not related to the high risk safety concerns covered by critical control points. Most importantly, the use of control points, as defined in this chapter, allows the food processor to separate the control of low probability safety risks from the significant safety risks covered by HACCP.

References

American Butter Institute/National Cheese Institute, (ABI/NCI). 1990. *Total Quality Systems Handbook—HACCP*. (Ed.) R.H. Ellinger. ABI/NCI, Washington, DC.

Bryan, F.L. 1990. Application of HACCP to ready-to-eat chilled foods. *Food Technol.* 45(7): 70–77.

Chilled Foods Association, (CFA). 1990. *Technical Handbook for the Chilled Foods Industry*. CFA, Atlanta, GA.

Corlett, D.A. and Stier, R.F. 1991. Risk assessment within the HACCP system. *Food Control*. 2: 71–72.

National Archives and Records Administration (NARA). 1990. *Code of Federal Regulations*. NARA, 21 CFR Parts 100 to 169. U.S. Government Printing Office, Washington, DC.

National Food Processors Association, (NFPA). 1989. *Guidelines for the Development, Production, Distribution & Handling of Refrigerated Foods*. NFPA, Washington, DC.

International Commission on Microbiological Specifications for Foods. (ICMSF). 1988. *Microorganisms in Foods. 4. Application of the Hazard Analysis Critical Control Point System to Ensure Microbiological Food Safety and Quality*. Blackwell Scientific Publications, London.

National Academy of Services (NAS). 1985. *An Evaluation of the Role of Microbiological Criteria for Foods and Food Ingredients*. NAS. National Research Council. National Academy Press, Washington, DC.

National Advisory Committee on Microbiological Criteria for Foods (NACMCF). 1990. *HACCP Principles for Food Production*. USDA-FSIS Information Office, Washington DC.

Sperber, W.H. 1991. The modern HACCP system. *Food Technol.* 45(6): 116–120.

US Department of Agriculture (USDA). 1989. *A Margin of Safety: the HACCP Approach to Food Safety Education*. USDA-FSIS Washington, DC.

US Department of Agriculture, (USDA). 1990a. *Q&A—HACCP Questions and Answers—Part II*. USDA-FSIS, Washington, DC.

US Department of Agriculture, (USDA). 1990b. *Summary of Public Hearings and Written Comments on Hazard Analysis and Critical Control Point Systems*. USDA-FSIS, Washington, DC.

12

Putting the Pieces Together: Developing an Action Plan for Implementing HACCP

Donald A. Corlett Jr.

INTRODUCTION

Implementing the HACCP system in a food company or manufacturing facility requires knowledge of the system, commitment, planning, resources and follow-through. Although the knowledge and technology of the HACCP system are important prerequisites, implementation is accomplished by the more conventional methods that organizations use for establishing new procedures and processes. One must not overlook the fact that people establish new systems and that organizational skills are a most important element of this type of endeavor. The following parts of this chapter discuss organizational elements of implementing HACCP.

WHERE DOES HACCP FIT INTO A COMPANY?

HACCP must be an integral part of the company quality assurance program. It is the cornerstone of the company's product safety system and fits into the umbrella quality assurance program as illustrated in Table 12-1. Note in the table that product safety is mandatory and that HACCP clearly belongs under this heading. Some required regulatory compliance matters involve product safety

TABLE 12-1   Umbrella Company Quality Assurance Program

	Product Safety	Regulatory Compliance	Product Quality
Status	Mandatory	Required	Voluntary
System	HACCP	Legal compliance	Quality control
Type of control point required	Critical Control point (CCP)	Regulatory control point (RCP)	Control point (CP)

Source: Corlett, D.A. 1991a.

and belong in the safety *and* regulatory systems. An example is pH control for an acidified low-acid canned food. However, not all regulatory compliance matters are critical to food safety and these do not belong in the HACCP system. Product quality is important, but is a negotiable matter that is voluntary and should be covered in the quality control system.

In Table 12-1, also note that product safety and the HACCP system require application of critical control points (CCPs). As noted in previous chapters of this volume, CCPs necessitate very heavy monitoring because they are used for preventing or controlling potential food safety hazards. When a CCP is determined to be out of control, mandatory action must be instituted to place affected product on-hold and take immediate correction action in the production system. Regulatory control points (RCPs) and control points (CPs) are intended for regulatory or quality attributes that do not involve this intensive monitoring and action.

ESTABLISHING HACCP ACCOUNTABILITIES

Because HACCP is a fundamental part of the company's operating system, as well as a major investment in employee time and company resources, clear accountabilities need to be established. This also enhances management support and cooperation during implementation.

A suggested guide for establishing HACCP accountability is provided in Table 12-2. Management must clearly set the stride for development of the company HACCP system. Establishment of accountabilities is a good way to get started and guide primary and shared responsibilities for the HACCP activities. It is recognized that perhaps other groups than those indicated may be selected for

TABLE 12-2   Suggested Guide for Establishing Accountabilities for HACCP

HACCP Activity	Company Departments				
	Management	R&D	QA	Operations	Marketing
Policy	XX				
Objectives	XX	X	X	X	X
Develop procedures		XX	XX	X	
Approve procedures	XX	X	X	X	X
Implement		X	XX	X	
Operate		X	X	XX	
Revise		XX	X	X	X
Verify	X	X	XX		

XX = primary accountability; X = shared accountability.

Source: Corlett, D.A 1991a

some of the suggested accountabilities. *The important point is to assign specific HACCP accountabilities to responsible persons and groups.*

ESTABLISHING THE HACCP IMPLEMENTATION PROJECT PLAN

A suggested HACCP implementation project plan is provided in Fig. 12-1. It is similar in design to project plans typically used in many companies. The list of steps will need to be augmented with resource requirements and key individuals must be allocated time to implement the system. Provision should also be made to include research and testing needed for development of technical procedures, critical control points and their critical limits.

The first step in using the project plan is for management to take the action to approve and organize it. All subsequent steps are influenced by this first step which requires uncompromising and sustained backing from senior management. Efforts to implement HACCP at more junior management levels, without support from senior management, are doomed to failure. In fact, a poorly developed or supported HACCP program may give the company a false sense of security and lead to many problems, particularly if conventional controls are also reduced.

The two management actions necessary to initiate the HACCP Implementation Project Plan are to: (1) develop HACCP policy and objectives, and (2) appoint the Core HACCP Team Coordinator and the Core HACCP Team.

Form 12-1

HACCP Implementation Project Plan

Page 1 of 1

Company: _____

Facility: _____

Products: _____

Date: _____ Revision: _____ Revision: _____

Core HACCP Team Coordinator: _____

Core HACCP Team Members
(Name and Title)

Activity	Responsibility	Action Dates				
		Start	Progress Reports			Finish
			-1-	-2-	-3-	
1. Develop HACCP policy, objectives, and implementation schedule						
2. Appoint core team HACCP coordinator and core HACCP team						
3. Train core HACCP team to prepare HACCP plan & implement the HACCP system						
4. Prepare model HACCP plan for one key product						
5. Appoint product HACCP team leaders and teams						
6. Conduct training for product teams; Use model HACCP plan						
7. Product teams prepare HACCP plans for products						

--continued on next page

FIGURE 12-1.   Form 12-1. HACCP implementation project plan.

Form 12-1 (continued)

Page 2 of 2.

Company:

Facility:

Products:

Date: _____ Revision: _____ Revision: _____

Core HACCP Team Coordinator:

Activity	Responsivility	Action Dates				
		Start	Progress Reports			Finish
			-1-	-2-	-3-	
8. Develop supervisor & operator procedures						
9. Conduct trial test of HACCP system on one product						
10. Evaluate results of trial HACCP system and make necessary changes						
11. Provide training to all employees (1-3 hour overview)						
12. Implement HACCP system for all products						
13. Conduct HACCP verifications for each product	HACCP coord. from another facility, or corp. QA					
14. Up-date and revise HACCP plans and HACCP system	Core HACCP coord and team					

ESTABLISHING POLICY

A common obstacle in getting started is developing a company policy for HACCP. This is not difficult. A suggested policy and two related objectives are:

HACCP Policy: "All company products will be safe for consumption."
HACCP   1. "The company will plan and implement a HACCP
Objectives: system for product safety."
                   2. "The HACCP system will be operational by
                       _____. (A stated date.)

APPOINTMENT OF THE CORE HACCP COORDINATOR AND TEAM

The Core HACCP Team ideally consists of persons with the supervisory and technical skills necessary to implement a major project. The team should consist of persons having manufacturing, distribution, quality control, research and development, engineering and sanitation responsibilities. The Core HACCP Team Coordinator must have organizational skills, a knowledge of the manufacturing operations, and be conversant with the technical aspects of producing the product(s). Once the team is appointed and assembled, they begin to implement the project plan given in Fig. 12-1.

A final note is that the coordinator and team need to become familiar with the HACCP system before they begin their work in order to implement it correctly. Of equal importance, they will ultimately become the personnel to educate others in use of the system in the company or facility.

DEVELOPMENT OF THE PRODUCT SPECIFIC HACCP PLAN

The initial focus of the HACCP coordinator and his team is development of the product specific HACCP plans for food products or groups of related products. The *HACCP Plan* is defined as, "The written document which delineates the formal procedures to be followed in accordance with these general principles." (NACMCF 1990)

Suggested guidelines for elements of the HACCP plan are given in Table 12-3. The HACCP plan is the blueprint for the company product-specific food plant HACCP system.

TABLE 12-3   Suggested Guidelines for Elements of the HACCP Plan

1. Designate a person responsible for the HACCP plan, and members of the HACCP "team" for the food facility and target product(s).
2. Organize the HACCP product safety system within and as part of the company quality assurance policy and program.
3. List the target food products, describe each product, list raw materials and ingredients, and prepare a preliminary flow diagram.
4. Document the hazard analysis and risk categories associated with the target products, their ingredients, and for the hazards in the entire product food chain (Principle #1).
5. Develop individual flow diagrams for each product that document the location and type of critical control points (CCP) for identified hazards (Principle #2).
6. Document description of each CCP, including the type of hazard, procedures or processes to control the hazard, and definition of the critical limits or tolerances that apply to each CCP (Principles #2 and #3).
7. Document monitoring procedures for the CCP and critical limits, monitoring frequency, and person(s) responsible for specific monitoring activities (Principle #4).
8. Document deviation procedures for each CCP, that specify action to be taken if monitoring determines that a CCP is out of control. Action must include safe disposition of affected food and correction of procedures or conditions that caused the out-of-control situation (Principle #5).
9. Develop and document record-keeping systems for the HACCP system using Principle #6. Designate trained and responsible company personnel for management and sign-off of records, and provide for record sign-off by a responsible official of the company.
10. Develop and document verification procedures based on Principle #7. Designate responsible company personnel to conduct verification of compliance to the HACCP plan and system on a scheduled basis. Designate responsible persons to conduct verification who are not generally involved in the line HACCP functions (such as Corporate or Division Quality Assurance).
11. Document procedures for revision and updating of the HACCP plan any time there is a change of ingredients, products, manufacturing conditions, evidence of new potential or actual hazard risks, or any other reasons that may influence the safety of the product(s). Otherwise specify scheduled revision and updating.
12. Consult with appropriate regulatory agency(s) regarding company intention to develop a HACCP plan, and involve the agency in development and approval of HACCP plan.

Source: Corlett, D.A. 1991b.

DEVELOPING PRODUCTION SUPERVISOR OR INDIVIDUAL OPERATOR PROCEDURES

Figure 12-2, "HACCP Production Supervisor or Individual Operator Procedures," is useful for developing actual procedures on the line, and is an aid to activity 6 in the HACCP implementation Projection Plan (Fig. 12-1). This form is included because one of the most challenging parts of developing a HACCP system is to "translate" monitoring and corrective action requirements into practical operating procedures for employees to use.

Form 12-2

HACCP Production Supervisor or Individual Operator Procedures:
Page 1 of 2 for this Worksheet Form:

Product:	Shift:
	Location:
CCP Number:	Supervisor:
Hazard Controlled:	Operator:

Critical Limits		Monitoring Inspection			Corrective Action	
#	Description	Activity	Frequency	Specification	Line	Product

Guide to Preparation of Procedures for the Above Critical Control Point:

How is the monitoring inspection conducted? Must equipment be dismantled? What is looked for? How will the inspection person determine if the critical limit is in specification? How would you tell a person to monitor this or these critical limits?

INSTRUCTIONS:

--Continued on next page

FIGURE 12-2.   Form 12-2. HACCP production supervisor or individual operator procedures.

Form 12-2

Page 2 of 2.
Continuation: HACCP Production Supervisor or Individual Operator Procedure

Product:	Shift:
	Location:
CCP Number:	Supervisor:
Hazard Controlled:	Operator:

What corrective action must be taken if monitoring indicates that the CCP is out of control? Who does the operator tell? When does he tell the person when the CCP is out of Control? What action does he take? How would you tell a person to take corrective action? Instruction:

1. Immediate corrective action for the manufacturing line:

2. Who does person notify and when?

3. Immediate action to place product on-hold:

4. Who does person notify that product is on-hold and when?

OPERPRO2   ©1991 Corlett Food Consulting Service. All Rights Reserved.

GETTING THE JOB COMPLETED

Once the HACCP project plan is initiated by management, the appointed co-ordinator and his team start the ball rolling. In the author's experience, the combination of management action to initiate HACCP and the appointment of the HACCP coordinator and the team, is a very effective way to implement the HACCP system.

Product safety and the HACCP system are team efforts. Everyone is responsible for food product safety and care must be taken to involve all employees in a production facility in development and training for using the HACCP system. All employees must understand that they play a most important role in using the HACCP system, because people make food safe to eat. Employees should be encouraged to be a part of the HACCP team and offer their views and expertise to continuously improve the facility HACCP system.

A carefully organized HACCP program involving all employees creates attitudes of awareness and action to prevent product safety hazards.

References

Corlett, D.A. 1991a. Initiating HACCP in your food company by establishing account-abilities, goals and a project plan. From course manual: *A Practical Application of HACCP*. Corlett Food Consulting Services, D.A. Corlett Jr., 5745 Amaranth Place, Concord, CA 94521.

Corlett, D.A. 1991b. Regulatory verification of HACCP systems. Food Technol. 45(4): 144.

National Advisory Committee on Microbiological Criteria for Foods (NACMCF). 1990. *HACCP Principles for Food Production*. USDA-FSIS Information Office, Washington, DC.

13

HACCP System in Regulatory Inspection Programs: Case Studies of the USDA, USDC, and DOD

Catherine E. Adams, E. Spencer Garrett,
Martha Hudak-Roos, E. Jeff Rhodehamel
and Dale D. Boyle

INTRODUCTION

The HACCP system has been utilized by certain food companies for several years. The system's application has been limited in the public sector; however, many expert scientific bodies promoted its use as a tool for regulatory programs to control food safety. The National Academy of Sciences (NAS 1985) included strong recommendations for HACCP application in regulatory programs in its 1985 report entitled, "An Evaluation of the Role of Microbiological Criteria for Foods and Food Ingredients." The NAS was requested in 1980 by four federal regulatory agencies to examine the potential applications for microbiological criteria in foods. The recommendation that resulted was to apply HACCP as an optimal system for preventing food safety problems. Specifically, the NAS report (NAS 1985) stated that HACCP *provides a more specific and critical approach to the control of microbiological hazards in foods than that provided by traditional inspection and quality control approaches . . .*"

The NAS Subcommittee which generated the report recognized that HACCP, as part of a food inspection program, required, (1) that inspection be focused more on monitoring than on finished product testing and (2) cooperation between government regulators and the regulated industry. The elements they identified as key to the success of a HACCP application included: (1) government and industry working cooperatively; (2) required training of industry and in-plant personnel; (3) parallel required training of inspectors; and (4) use of the HACCP system mandated by federal regulation.

The NAS identified HACCP as a technically sophisticated system requiring considerable preparation before application. However, the NAS report made clear the potential connection between the successful use of HACCP in food protection systems and potential meaningful reductions in the incidence of foodborne disease in the United States.

The intent of regulatory agencies to reduce foodborne disease prompted officials at the U.S. Department of Agriculture, Food Safety and Inspection Service (USDA-FSIS) and U.S. Department of Commerce (USDC), National Marine Fisheries Service (NMFS) to develop studies to apply HACCP as a tool for inspection. The following sections detail the process used to design implementation programs for HACCP to meat and poultry inspection by FSIS and for fish and seafood inspection by NMFS. Distinctions between the two approaches in terms of focus on safety vs safety and quality/economic aspects stem from major differences in the current regulatory status for the commodities. Mandatory inspection has been in effect for meat and poultry products for many years; and while fish and seafood inspection has some mandatory inspection, NMFS's inspection is voluntary.

The Department of Defense (DOD) is currently using HACCP for some food procurement and handling programs. From the inception of HACCP application, DOD has been actively involved. The U.S. Army Natick Research, Development & Engineering Center (NRDEC) cooperated in the development of HACCP applications for food. The application of HACCP has now moved into retail commissary operations.

USDA-FSIS HACCP PROJECT

In August, 1989, the Administrator of FSIS announced their intent to apply HACCP to meat and poultry inspection as part of inspection modernization. The Agency distributed a Concept Paper (USDA-FSIS 1989) outlining a four-step study to determine the optimal process for HACCP implementation. The intent was to systematically develop a HACCP-based inspection program, with cooperation from all involved parties. The HACCP study included:

(1) solicitation from employees, employee organizations (including the labor union, consumer representative groups, and industry, which resulted in over 100 meetings with over 3000 people) and five public hearings;
(2) workshops with industry technical experts facilitated by Agency employees to determine generic HACCP plans for select categories of products;

(3) in-plant testing of the generic HACCP plans; and

(4) application of predetermined evaluation criteria.

Specific details of the HACCP study were presented in a strategy paper (USDA-FSIS 1990a) distributed in January, 1990.

The plan was to base HACCP implementation by FSIS on the outcome of the HACCP study. The HACCP study and the evaluation criteria were peer reviewed by a team of experts, representing fields including public health, food science, veterinary medicine, statistics, and quality management systems. The Evaluation Plan (FSIS 1991) was distributed to all interested parties, including all organizations included in the original solicitation for ideas.

FSIS initially intended to include the full scope of its regulatory responsibilities in the HACCP study, i.e., safety, wholesomeness, and prevention of economic adulteration. During the solicitation and public hearing process, it was clear there was consensus regarding restriction of HACCP application to food safety issues only. Information and opinions were collected from over 100 organizations (USDA-FSIS 1990b). There was universal concern among the industry, employee and other professional organizations, the scientific community, employees, and consumer representative groups that the safety control of a HACCP-based inspection system would be reduced if critical control points (CCPs) were permitted to address nonsafety-related issues. One purpose for the Agency's decision to solicit information from the public was to develop a consensus for and commitment to HACCP implementation. Therefore, the Agency listened to concerns on behalf of all interest groups and decided to consider only safety issues as critical.

FSIS appointed a Core Team to direct all HACCP functions. Because the Agency was cutting across disciplinary lines to accomplish the goals of the HACCP study, senior-level officials were selected for the five-member HACCP Core Team. The Core Team selected, through a competitive process, a six-member Special Team and a Director. Only the Director's position was full time on the HACCP study. The Special Team members were one-half time appointments. Special Team members were selected on the basis of their field experience, knowledge of meat and poultry inspection regulation, ability to manage complex assignments, and recognized ability as leaders. All but one member were field-based, with one from Washington, DC Headquarters office. Individuals were selected representing professions as veterinarians and food scientists. Knowledge and experience in slaughter, processing, and import inspections were sought.

Once selected, the team underwent thorough training in HACCP systems, quality management systems, microbiology, and group facilitation. They met frequently and learned to work closely as a team. In facilitating meetings, they learned to develop consensus and a feeling of ownership for the work product, the HACCP plan, on behalf of their participants. This sense of ownership on behalf of the industry participants in the workshops was vital to complying with the NAS recommendation to develop commitment by the industry for HACCP operations.

In February, 1991, the first of the industry/FSIS workshops was held. Each of five workshops was to have approximately 40 industry volunteers as participants. The workshops were held on five topics representing the recommendations heard during the solicitation phase of the HACCP study. These topics included:

(1) refrigerated prepared foods (cook-and-assemble style)
(2) cooked sausage
(3) poultry slaughter
(4) fresh ground beef and patty manufacture, and
(5) swine slaughter.

The topics were selected based on consensus from various groups and diversity of scope. Products were selected to include slaughter operations, cooked ready-to-eat products, and raw products that were minimally processed but not cooked.

Led by the FSIS Special Team facilitators, the groups developed a generic HACCP plan for each category. CCPs were identified by following procedures outlined by the National Advisory Committee for Microbiological Criteria for Foods (NACMCF 1989). When appropriate, the logic sequence and decision tree approach being developed by the Codex Alimentarius Commission's Food Hygiene Committee (Appendix B) were utilized. In identifying CCPs, technical experts were able to draw on data collected by their companies or by trade organizations. If recommendations deviated from current USDA regulations, scientific documentation for the deviation could be presented to initiate a change in the regulation.

Subject Matter Experts (SMEs) were incorporated into workshops to represent USDA inspectors from the field. Individuals were selected competitively through the regions. Current experience as a USDA inspector in the type of plant being considered was a critical criterion for selection. The involvement of these skilled and committed individuals proved to be an invaluable part of the HACCP study.

The second phase of the HACCP study included in-plant testing of the generic HACCP plans developed in the workshops. Three volunteering plants were selected for each category of products. The plants were selected based on representation of small, medium, and large operations. The intent of this scope of plant size and complexity was to demonstrate HACCP's effectiveness as an inspection tool for all sizes of operations.

For in-plant tests, the generic HACCP plans were tailored to fit the particular product and conditions existing in the plant. Plant management was responsible for the tailoring of the generic HACCP plan to fit their plant. FSIS Special Team members trained plant employees and inspection employees on HACCP application.

After in-plant testing had begun, data were collected for evaluation of HACCP effectiveness as an inspection system. Evaluation criteria were peer reviewed and focused on all aspects of food safety, including microbiological, physical, and chemical potential hazards. Evaluation included sampling and analytical tests as well as monitoring reports. Along with quantitative data from sampling, qualitative data were collected from inspection and plant employees. These subjective data augment the objective information and recognize that people's perception is an important part of a successful food inspection program.

At the conclusion of the 30-month study, recommendations regarding the optimal process for implementation of a HACCP-based meat and poultry inspection program will be forwarded by the peer review panel, the HACCP Special Team and the Core Team to the FSIS Administrator. The results of the study should indicate whether the process of open workshops for technical experts of the industry is a viable one for advancing HACCP implementation for FSIS. The process used for implementing HACCP should, however, be a deliberate one which slowly incorporates all regulated products. The process should be considered "evolutionary," and not "revolutionary".

HACCP application by FSIS was designed to capture the elements determined to be important for successful implementation by NAS and also elements of quality management systems. HACCP may be considered a food safety program within a quality management program. The systems and people orientation of the two respective programs are paramount to success. The need for commitment from top management and down the chain of command is critical for HACCP or quality programs to work. There is also need for adequately trained staff to operate the system effectively and recognize that the operation is data driven.

USDA's FSIS effectively captured these elements within the first 18 months of the HACCP study. The industry was a willing and committed partner in HACCP development. The Agency's employees were aware of the HACCP study and were supportive. FSIS's labor organization and other professional organizations spoke highly of the effort in public—strong endorsements of support at mid-term of the agency's study to implement HACCP as an inspection tool.

The project is still active and scheduled for completion with evaluation in late 1992. Plans should then be announced for use of HACCP in meat and poultry inspection.

NMFS HACCP PROJECT

NMFS began its examination of applying HACCP to the seafood industry in 1986 when Congress requested the National Oceanic and Atmospheric Administration (NOAA) to design "a program of certification and surveillance to improve the inspection of fish and seafood consistent with the Hazard Analysis and

Critical Control Point system" (U.S. Congress 1986). NOAA/NMFS called the study the Model Seafood Surveillance Project (MSSP). The study classified hazards in the consumption of seafood by consumers into three categories: product safety, plant/food hygiene (wholesomeness and sanitation), and economic fraud (NMFS 1987). These classifications were then compared to traditional HACCP theory and applications.

It was determined that there were impediments to applying traditional HACCP concepts to the seafood industry dominated by marketing of fresh and frozen products. Traditional and accepted HACCP applications to only product safety concerns omitted other significant concerns for fresh and frozen seafood products (i.e. decomposition and wholesomeness), in a relatively minimally regulated marketplace.

Second, the early work in HACCP was done primarily by scientists who were mainly concerned about product safety, especially as it related to microbiology. Restricting the concept to solely microbiological product safety issues was considered inappropriate for the MSSP because seafood product safety concerns include a number of chemical toxins in addition to microbiological pathogens.

Finally, the seafood industry had expressed concerns regarding the extent of economic fraud in the industry (NFI 1985). A workable program that would control all potential hazards was desirable as NMFS worked to incorporate HACCP-based controls for this relatively minimally regulated industry.

Therefore, during the MSSP study and with seafood industry's concurrence, it was decided to develop a regulatory program to cover all potential consumer hazards: product safety, plant/food hygiene, and economic fraud. The MSSP accomplished this task through a series of workshops, site testing, and industry meetings. NMFS and NFI conducted 49 workshops with over 1200 participants and 280 site tests with 20 industry steering committee meetings. Industry evaluations and study results indicated that the comprehensive expanded HACCP program was feasible and desirable from the seafood industry's and NMFS's regulatory perspectives.

HACCP is a system of control and NMFS found it could be used as an "umbrella" approach to control food processes. By applying HACCP to safety, wholesomeness, and prevention of economic fraud, one focused system was created that pinpoints areas in the process that must be controlled to prevent a consumer hazard and ultimately, a recall. A recall of 40,000 pounds of product would cost a firm the same amount regardless of whether the cause is safety, adulteration, or labeling.

The MSSP study determined CCPs for the seafood industry's HACCP applications. For the majority of seafood commodities (even the most complex processing procedures), product safety, plant/food hygiene, and economic adulteration could all be controlled with 11 to 38 CCPs.

The seafood industry found it useful to include sanitation and economic fraud in their training for a HACCP approach to a regulatory program. By treating all CCPs the same, they found there was less confusion during training and employees could understand that "this point is critical because it resulted in a product that was not in compliance with the law," (i.e., not safe, unwholesome, or mislabeled). This was especially true in small plants where no formal quality control or quality assurance department existed and a limited number of people were responsible for monitoring and verifying CCPs.

The MSSP study and the pilot NOAA program have proven that regulatory records could be generated that met all requirements but did not expose proprietary information. The records must demonstrate that control was maintained at CCPs. If control was lost, the records must indicate corrective actions and proper disposition of the product. In the seafood industry, records design or forms used were left up to the individual firm, with approval by the regulatory bodies. This allowance gave the firm the ability to limit proprietary information on regulatory records.

The vast majority of the seafood industry supported the expansion of HACCP and some plants have chosen to apply it in a voluntary regulatory program sponsored by the Food & Drug Administration (FDA) and NMFS. Plants participating in this program were enthusiastic and convinced that HACCP works. These include small plants with less than 100 employees that produce less than 1 million pounds of product per year.

JOINT FDA/NOAA SEAFOOD INSPECTION PROGRAM

The Food & Drug Administration (FDA) and National Oceanic and Atmospheric Administration (NOAA) have developed a new voluntary inspection program, to be jointly operated, that will build upon the current resources and expertise of both agencies (FDA/NOAA 1990).

This program is based upon the HACCP concept that is expanded to include economic and sanitation parameters. It is a fee-for-service inspection program that will use an official mark to indicate the product is from a facility meeting the requirements of the program. The program covers seafood from water to table including retail establishments.

Examples of products that may be included under this program are: low-acid or acidified canned fish or fishery products; cooked, ready-to-eat refrigerated or frozen fish and fishery products; fresh fish, whole or gutted; or crustaceans that are refrigerated or frozen; fillets of fish; peeled and deveined crustaceans that require further heat processing; imitation fishery products; in-shell or shucked molluscan shellfish; and fish meal and by-products for use as animal feed.

Seafood is regulated at the Federal level primarily by the FDA, under the authorities of the Federal Food, Drug, and Cosmetic (FD&C) Act and the Public

Health Service (PHS) Act. The FD&C Act charges the FDA with assuring that foods, including seafood, are safe, wholesome, and not misbranded or deceptively packaged. FDA's authority under the PHS Act relates to the control of communicable diseases from one state, territory, or possession to another, or from outside the United States into this country.

NOAA operates a voluntary, fee-for-service seafood inspection program that is conducted under the authorities of the Agriculture Marketing Act of 1946 and the Fisheries and Wildlife Act (1956). A Memorandum of Understanding (MOU) exists between NOAA and FDA. As part of the MOU, NOAA ensures that its client's operations and products meet the requirements of the FD&C Act as well as NOAA's own quality and identity requirements. The program includes inspection, grading, and certification services, as well as the use of official marks that indicate that a client's products have been federally inspected.

FDA and NOAA expect that a joint HACCP-based program will lead to more efficient regulation of the seafood industry and add a further assurance of safety, wholesomeness and truthful labeling over that which already exists. It will also provide recognition to the industry for successfully operating under a HACCP-based program.

The HACCP-based program emphasizes the industry's role in continuous problem prevention and problem solving from the water to the consumer for both human and animal foods. Relying on periodic facility inspections and analysis of end product samples by the government is not desirable as sole means for ensuring food safety and compliance with the law.

In formulating the joint program, FDA and NOAA followed the principle recommendations of the November, 1985 National Academy of Sciences (NAS 1985) report; considered the recommendations of the 1989, National Advisory Committee on Microbiological Criteria for Foods (NACMCF 1989); considered information provided from the NMFS Model Seafood Surveillance Project; considered information provided from the Quality Management Program for the Canadian Fish Processing Industry; and used both agencies' expertise in the application of HACCP procedures as well as the regulations promulgated by the FDA and NOAA.

It is important to note, however, that the voluntary program is not a self-certification program by the industry. While HACCP involves substantial self-monitoring of critical control points by the industry, to assure that program is effective, it will include regular monitoring inspections and less frequent verification inspections by the two agencies. By means of these inspections, FDA and NOAA will be able to determine whether each HACCP-based system is in compliance with a facility's HACCP plan, which includes checks for overall sanitation and compliance with good manufacturing practices, labeling, and other requirements.

Each facility or firm participating in the Joint Voluntary Program will be audited by local, state, or federal agencies through inspection of the facility and records review. Which agency performs the audit will depend upon the particular segment of the program (retail, processors, vessels, etc.) or the location of the firm. In addition, sample collection and analyses may be used to determine compliance with the HACCP-based plan(s). If the responsible agency performing the audit deems that sample analyses is appropriate, then samples will be obtained from the facility cost-free for analysis. Reimbursement to the regulatory agency for the cost of analyses will be made by the firm.

As part of the auditing procedure, a "rating" and "deficiency" matrix will be applied to the HACCP-based plan to determine severity or defects (i.e., Minor, Major, Serious or Critical). The rating received by each facility (i.e., A, B, C, or D) will depend upon the number and severity of defects found and will be for internal use only. The rating will determine the frequency of the audits of the facility by the responsible local, state or federal agency and will vary with each segment of the Joint Program (retail, processors, vessels, etc.).

Verification of the Joint HACCP-based Program is under the purview of the FDA. During the verification the FDA will ascertain: (1) validity of the plan; (2) the facility's adherence to its HACCP-based plan; (3) proper government auditing procedures of the HACCP-based plan have been applied; and (4) adherence to the requirements of the FD&C Act.

HACCP APPLICATIONS IN FOOD
PROCUREMENT AND HANDLING FOR DOD

Within DOD food management systems, HACCP is actively being applied. Retail food commissaries compose a major responsibility for the U.S. Army Veterinary Corps. One commissary, located at Fort Meyer, Virginia, is already using a HACCP approach for procurement and handling. HACCP will be a part of the new Defense Commissary agency's management strategy. Inspectors will validate the effectiveness of the system in the future with each commissary responsible for their own HACCP operation to assure food safety.

With HACCP, inspectors assume the role of technical assistant and advisor rather than policemen. HACCP concepts are being taught to all Veterinary Corps officers, warrant officers, and all food inspection specialists at the Academy of Health Sciences, located at Fort Sam Houston, Texas. The DOD works cooperatively with Federal agencies, including USDA, FDA, and NMFS to ensure uniformity in HACCP applications.

Following the key role U.S. Army NRDEC played with respect to HACCP's original development, Natick is working on application for HACCP in hospital kitchens preparing food for critically ill and immunocompromised patients. The Defense Personnel Support Center requires strict quality control procedures similar to HACCP for producers and assemblers of military food rations. Approved statistical process control plans are now required before a vendor is allowed to start manufacturing food for military personnel. The experience has already shown a reduction in expensive reworks and recalls for government contractors.

The military's use of "Best Value" buying techniques should greatly encourage the industry's adoption and use of HACCP systems. "Best Value Procurement" is being used by DOD to purchase foods. This system allows the government to buy the best available product for a given price. No longer are low bidders assured being awarded a contract for products. The potential contractor is evaluated on the merits of previous performance, quality assurance, and ability to produce the desired quantity of product on time and at the stated price. A working HACCP program could be greatly beneficial in being awarded government contracts today and in the future.

For DOD, viable HACCP systems have the benefit of assuring safe foods, with a secondary benefit of enhanced quality.

CONCLUSIONS

Government agencies are actively seeking opportunities to implement HACCP systems for food safety management. Use of HACCP systems will better ensure the prevention of manufacturing problems relative to any known quality control system. Projects currently in place within government agencies should verify HACCP's use for food safety and encourage a wide sector of the food industry to implement HACCP systems.

The issue regarding scope of HACCP applications—safety only vs. safety, hygiene, and prevention of economic adulteration—remains a topic of debate. The answer may lie in the realization that different segments of the food industry have differing needs, and one type of application may not fit all needs. Current regulations differ for various food commodities. These different regulatory concerns are not an issue which should affect the success of HACCP systems in operation.

HACCP applications provide a useful and beneficial tool for all elements of the food industry, and can be a useful tool for government regulators in inspection programs. HACCP can create greater levels of cooperation between government and industry, thereby better accomplishing the mutual goal to prepare and market safe foods for American consumers.

References

U.S. Food and Drug Administration and National Oceanic and Atmospheric Administration (FDA and NOAA). 1990. FDA/NOAA Joint Program Submission Guide. FDA, Office of Seafood, Washington, DC.

National Advisory Committee for Microbiological Criteria for Foods (NACMCF). 1989. HACCP Principles for Food Production NACMCF. USDA-FSIS Information Office Washington DC.

National Academy of Sciences (NAS). 1985. *An Evaluation of the Role of Microbiological Criteria for Foods and Food Ingredients*. NAS, National Research Council, National Academy Press, Washington, DC.

National Fisheries Institute (NFI). 1985. Seafood Quality: Inspecting the Issue: A study of the U.S. Fisheries Inspection system conducted by the National Fisheries Institute in cooperation with National Marine Fisheries Service. NFI, Washington, DC.

National Marine Fisheries Service (NMFS). 1987. Plan of Operation—Model Seafood Surveillance Project. NMFS, Office of Trade and Industry Services, National Seafood Inspection Laboratory, Pascagoula, MS.

U.S. Congress. 1986. Senate Appropriations Committee: Report to accompany H.R. 5161. Rept. 99-425, 99th Congress, 2nd Session, Sept. 3.

U.S. Department of Agriculture-Food Safety and Inspection Service (USDA-FSIS). 1989. Hazard Analysis and Critical Control Point Applications for Meat and Poultry Inspection: A Concept Paper. USDA-FSIS, Information Office, Washington, DC.

U.S. Department of Agriculture-Food Safety and Inspection Service (USDA-FSIS). 1990a. Hazard Analysis and Critical Control Point Application in Meat and Poultry Inspection: A Strategy Paper. USDA-FSIS Information Office, Washington, DC, January.

U.S. Department of Agriculture-Food Safety and Inspection Service (USDA-FSIS). 1990b. Summary of Public Hearings and Written Comments on Hazard Analysis and Critical Control Point System (HACCP). Policy Analysis Unit, USDA-FSIS Information Office, Washington, DC.

U.S. Department of Agriculture-Food Safety and Inspection Service (USDA-FSIS). 1991. Evaluation Plan for the HACCP Study. USDA-FSIS Information Office, Washington, DC.

14

Practical Application of HACCP

Richard Stier

INTRODUCTION

Many companies are in the process of initiating HACCP programs, are considering it, and/or requiring their suppliers to do so. Implementation of a HACCP program can not be accomplished overnight. Depending upon the degree of sophistication inherent in a company's operations, it may take anywhere from six months to two–three years. Getting started with HACCP may be the most difficult part of the program, followed closely by putting a plan into practice. Most technical staff recognize how vastly different theory and practice are. The bench-top formulation always requires modification when put into production; pilot plant scale studies can provide insight on how a product or process will perform, but do not mirror precisely what happens during a full-scale operation; and plans formulated around the conference table require alteration when put into practice. The same is true when a HACCP program is developed and implemented.

This chapter was developed to provide additional insight for those individuals planning to implement HACCP in their operations. The focus will include assessing risk, establishing critical control points, and setting limits. Different process models, based on actual production operations, are used. The hope is that one or more of these models will help readers in developing their own programs.

PRODUCT DESCRIPTION

For each model, a description of the product being processed will be provided. These descriptions may be used as guides for potential HACCP users. The description will include all applicable information on the product which will help in assessing risk and establishing Critical Control Points (CCPs). Information typically included in such descriptions are product name, target market, how the product will be used, the type of packaging, label instructions, an ingredients list, shelf life, and handling requirements (Corlett and Stier 1990).

RISK ASSESSMENT

For each of the models used, the risk on the product and its ingredients will be ascertained using established criteria for assessing microbiological risk (NACMCF 1989) developed by the NACMCF. Physical and chemical risk will be determined using procedures proposed by Corlett and Stier (1991a and b). Forms described by Corlett and Stier (1990) are used in this part of the model. The risk values stated for each product and component ingredients are not "carved in stone". They may vary for an individual material, depending upon source, production environment, and personal experience.

PROCESS FLOW CHARTS

To properly develop and implement a HACCP system, it is essential that individuals understand the process. Creation of accurate process flow charts is one step in developing understanding. Once the chart has been developed, and the risk assessments are completed, the "HACCP Team" may begin to assign Critical Control Points. One school of thought for instruction of HACCP principles involves a similar process; that is, presenting a flow chart to a class of students and allowing them to set CCPs (Corlett and Stier 1990). Use of these "naked flow charts" is an excellent teaching tool. In these models only flow charts with CCPs already assigned are employed.

CRITICAL CONTROL POINTS AND LIMITS

Critical Control Points will be established for each model. These Critical Control Points (CCPs) will be set which ensure the safety of the product in mind. Those points in the process flow which relate to quality, fraud, or aesthetics will not be deemed CPPs, unless product safety is affected. Establishment of CCPs has been and will remain a point for discussion when companies build their own HACCP systems. Too many points would make the system unwieldy, and too few would not assure the safety of the food. It is, probably, better to err on the high side, however. If a point is established as a CCP and a review of the data indicates no problem, the system can be modified. If a point is added later, questions may be raised on the safety of food prior to adding the controls. One area which always prompts a discussion is product coding. Some believe that product coding is a good manufacturing practice, and should be considered a quality issue. Others feel that because proper coding is essential for tracking and recovering product, a code is necessary to food safety. In these models the latter philosophy will be applied.

Once CCPs are established, limits for each CCP must be established. These limits must be monitored and acted upon should the system get "out of control". The action or more precisely reaction to a deviation is one of the seven HACCP principles and a key to making HACCP workable. Guidelines on how limits are established for models will be provided. Actual numbers, that is, times, temperatures, concentrations of chemicals or metabolites, or schedules, will not be provided, since these could be construed as recommendations. The objective of these models is to provide guidelines, not recommendations. Values which users of HACCP apply to their systems are generally unique to the process.

Finally, to designate the kind of CCPs being established, the letters M, P, C, and S will be used, to designate Microbiological, Chemical, Physical, and Sanitation, respectively.

MODELS

The models to be used are based on actual food processing operations. The models used include:

1. Canned mushrooms
2. Shredded lettuce
3. French fries (Par-fries)
4. Chicken salad

There are other models available to those interested in developing HACCP systems, which include those developed by NOAA/NMFS and are being developed by USDA/FSIS. Some models published by NMFS include those for blue crab, breaded shrimp, raw shrimp, and breaded fish and specialty items (NOAA/NMFS 1989a, 1989b, 1990a, 1990b).

Finally, several products are utilized in restaurant and institutional foodservice operations. Insights on how final users can assure the safety of these products are presented.

Canned mushrooms

This model describes the HACCP program developed to reduce risks in mushrooms produced by the People's Republic of China. Canned mushrooms have long been viewed as a somewhat hazardous product because of concerns about botulism. Mushrooms are, in fact, one of the few commercial low-acid foods which have been found to contain botulinal toxin since the Low Acid Canned Food Regulations were passed (Reister 1974). Following these outbreaks, steps were taken by the industry to modify (Denny 1982) processes to prevent future occurrences.

In 1989, a new problem surfaced in canned mushrooms. Product manufactured by the People's Republic of China (PRC) was the source of staphylococcal enterotoxin, which resulted in several outbreaks and numerous illnesses (Anonymous 1989). These incidents resulted in the detention of mushrooms produced by the PRC (NFPA 1989). The cause of the problem was improper packaging and handling of the fresh mushrooms, which created an environment conducive to the outgrowth and toxin production by *S. aureus* (Hardt-English 1990). Sufficient enterotoxin was produced so that it was not destroyed by the thermal process given to the mushrooms. The problem was compounded by the poor distribution networks inherent in rural China (which included shipping the raw product to the plant without refrigeration) and the political economic system existing in China at the time.

The model was developed for the PRC although it could be applied in countries where conditions were similar. A number of the CCPs which this HACCP system included would *not* be present in facilities where mushrooms were produced and processed in a single complex, or in operations where growers are more localized. Perhaps the most unique feature of this particular HACCP model is that it was developed to control the development of enterotoxin in a fresh product and incorporated the control points necessary to assure the safe production of low-acid canned foods, as described in 21 CFR Part 113.

Product description. The product may be described as follows: Mushrooms (buttons and pieces and stems) are packed in welded cans manufactured on-site. Can sizes ranged from 4 oz (211 × 211) to 603 × 700. They are brined with a salt solution and seamed. Filling, check weighing, and retort basket loading and unloading are all hand operations. The facilities are not modern, employing large numbers of people, instead of equipment for most unit operations.

The mushrooms are produced by farmers located in areas around the plants, some at distances of 60 kilometers. Product is purchased only from growers using known strains of mushroom supplied by the government. The mushrooms are harvested, graded, and transported to the canneries in baskets without being cooled or refrigerated.

The vast majority of product packaged in these plants is for export. Institutional size containers of mushrooms are used by the United States foodservice industry, especially the pizza industry. There are no special handling requirements, other than that the product be handled to prevent damage and possible leakage.

Risk assessment. *Microbiological.* The two organisms of concern with this particular product are *S. aureus*, in particular its enterotoxin, and *C. botulinum*. Should conditions be allowed to exist which would support the growth and toxin production by *S. aureus*, the thermal process designed to destroy *C. botulinum* might not destroy the enterotoxin. With this in mind, apply the principles for assessing microbial risk to the finished product and the component ingredients, water and salt (Fig. 14-1).

The target market for this product is not a special population. The raw mushrooms and the product itself are definitely sensitive; there is a kill step, but it may be inadequate if enterotoxin is present. Since this is a canned product, there is little chance for contamination occurring between processing and packaging. The product is shipped via container ship so there is a possibility for abusive handling in distribution, which could result in post process contamination.

Salt is not a hazardous ingredient, but water can be, depending upon the source. In this case, the assumption is a chlorinated city water, which has relatively low risk.

Chemical and physical. The chemical and physical risk assessments for canned mushrooms and ingredients may be seen in Fig. 14-2 and 14-3. The major concerns in this area are the inadvertent introduction of potentially toxic mushrooms into the system and physical adulteration during handling because of the large number of unit operations in the mushroom process. Warehouse conditions and lack of attention to proper labeling of ingredients and other materials increased the risk of chemical contamination of the salt.

Food Item	Microbiological Hazard Characteristics Associated with the Food and It's Ingredients						Hazard Category
(1) Product	A High Risk Special Population	B Sensitive Ingredients	C No Kill-Step in Process	D Recontam. between Proc/Pack	E Abusive Handling Dist/Cons	F No Term. Heat Proc. by Consumer	
Mushrooms	(-)	+	+/-	(-)	+	(+)/(-)	II - IV
(2) Raw Mat's and Ing's							
Mushroom	(-)	(+)	(+)	(+)	(+)	(+)	V
Water	(-)	(+)	(+)/(-)	(-)	(-)	(-)	II/I
Salt	(-)	(-)	(-)	(-)	(-)	(-)	0

Notes: (1) As used by consumer; (2) As entering the food facility before preparation or processing.

FIGURE 14-1.   Microbiological risk assessment for canned mushrooms.

Food Item	Hazard Characteristics Known to be Associated with the Food and It's Ingredients						Hazard Category
(1) Product	A High Risk Special Population	B Ingredients Contain Hazard	C Not Re- moved in Manufact.	D Recontam. between Mfg/Pack.	E Contam. by Dist. or Cons.	F Consumer Cannot Detect/ Rem.	
Mushrooms	(-)	(+)	(+)	(-)	(-)	(+)	III
(2) Raw Mat's and Ing's							
Mushroom	(-)	(+)	(+)	(+)	(+)	(+)	V
Water	(-)	(-)	(-)	(+)	(+)	(+)	III
Salt	(-)	(+)	(+)	(-)	(-)	(+)	III

Notes: (1) As used by consumer; (2) As entering the food facility before preparation or processing.

FIGURE 14-2.   Assessment of chemical risk for canned mushrooms.

Critical control points and limits. The process flow chart for production of mushrooms is shown in Fig. 14-4. This chart includes all Critical Control Points developed to ensure the safety of canned Chinese mushrooms. For ease of discussion, the CCPs and limits are grouped according to the following areas: cultivation and harvest, transport, processing, and post-process handling.

Food Item	Hazard Characteristics Known to be Associated with the Food and It's Ingredients						Hazard Category
(1) Product	A High Risk Special Population	B Ingredients Contain Hazard	C Not Re- moved in Manufact.	D Recontam. between Mfg/Pack.	E Contam. by Dist. or Cons.	F Consumer Cannot Detect/ Rem.	
Mushrooms	(-)	(+)	(-)	(-)	(-)	(+)	II
(2) Raw Mat's and Ing's							
Mushroom	(-)	(+)	(-)	(-)	(+)	(-)	II
Water	(-)	(-)	(-)	(-)	(-)	(-)	0
Salt	(-)	(-)	(-)	(-)	(-)	(-)	0

Notes: (1) As used by consumer; (2) As entering the food facility before preparation or processing.

FIGURE 14-3.   Assessment of physical risk for canned mushrooms.

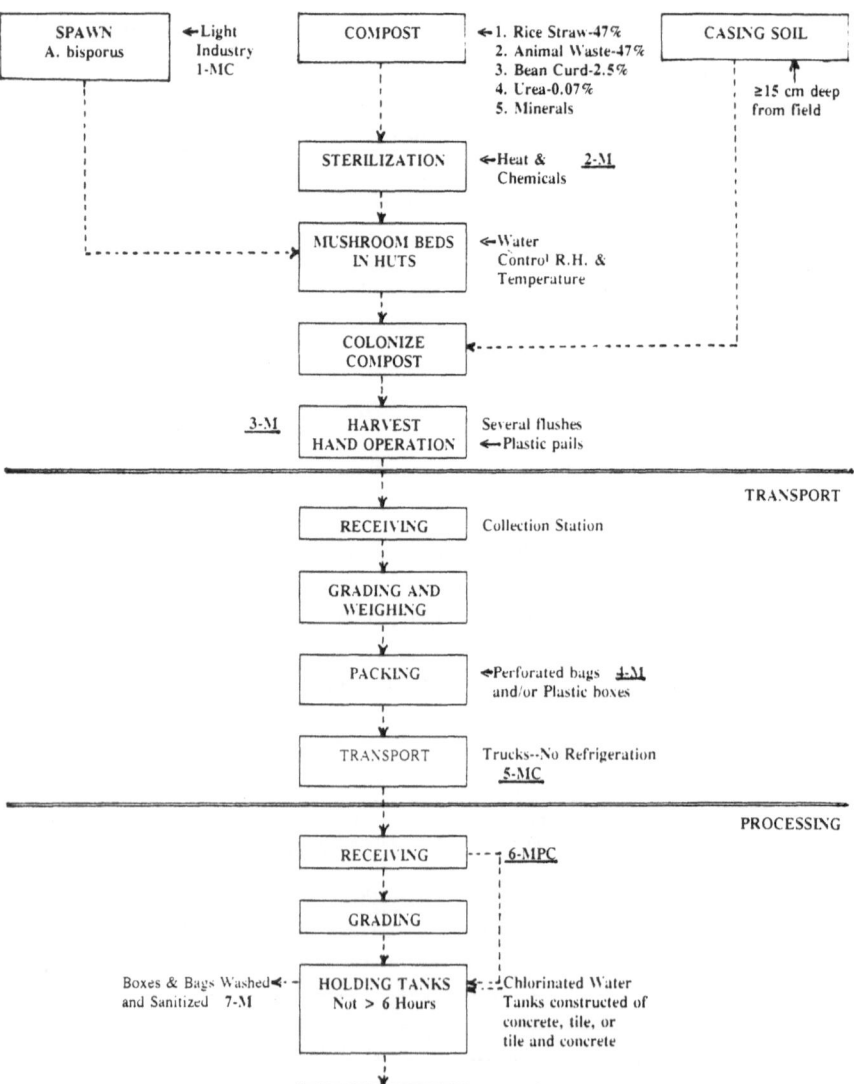

FIGURE 14-4. Mushroom production and processing (PRC). (M = Microbiological; C = Chemical; P = Physical; S = Sanitation).

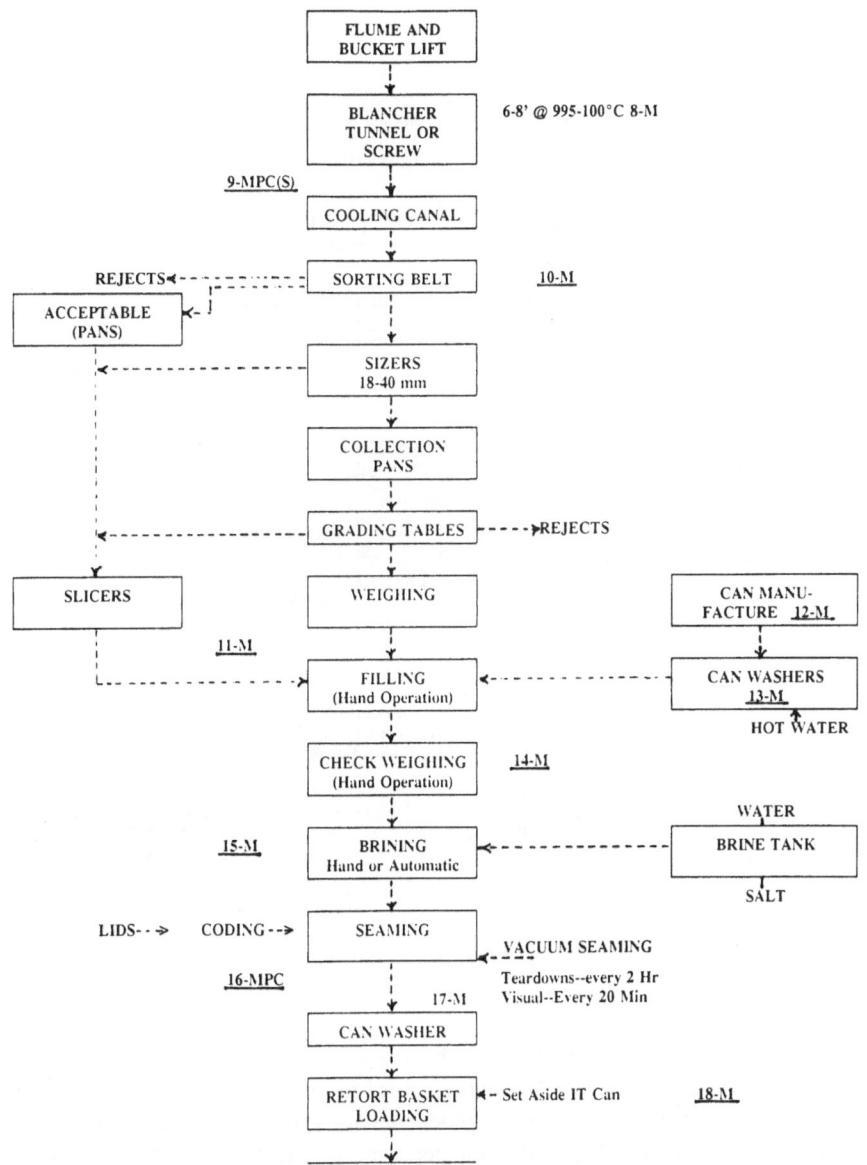

FIGURE 14-4.   Continued.

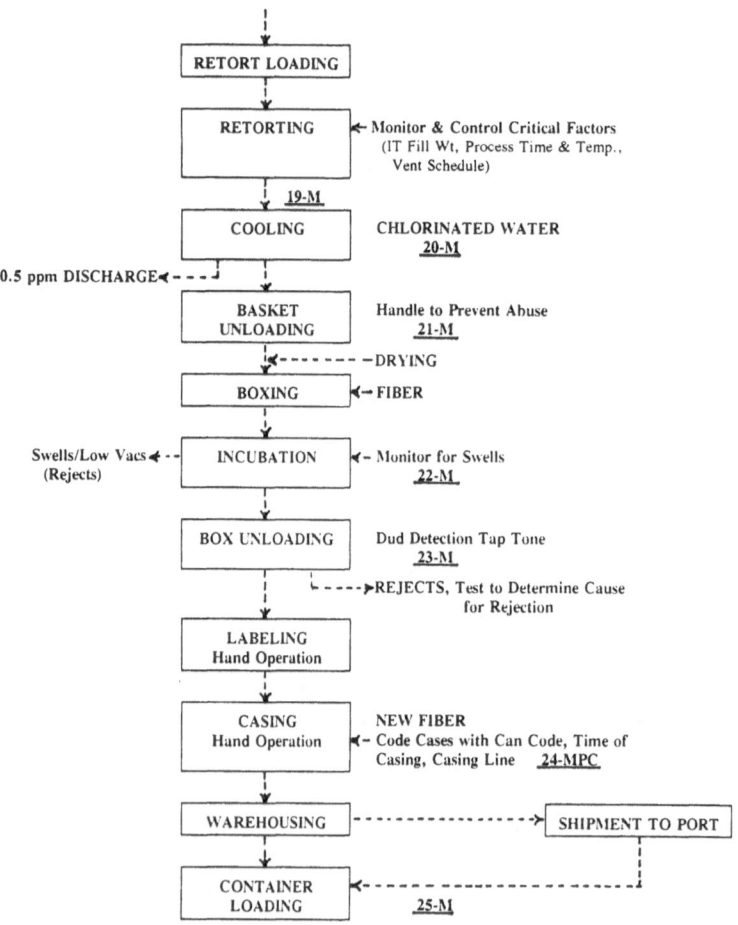

FIGURE 14-4.   Continued.

Cultivation and harvest. One of the areas which was thought a potential source for substandard and possibly toxic mushrooms was product obtained from what might best be described as "independent" growers. The issue has been addressed. The first CCP, 1-MC, creates what might be called in this country an "approved supplier." Only growers who receive certified spawn from the province's Light Industry group may sell to the canneries. This reduces the risk of allowing "wild" and possibly toxic strains into the system. The growers also receive training from Light Industry on proper cultivation, maintenance, and harvest techniques, all of which must be documented in actual practice.

CCP 2-M is a result of this program. Records must be maintained by each farmer for compost sterilization. Times and temperatures of heating and exposure to chemicals must be recorded. The final CCP (3-M) pertains to harvest. Farmers must follow set protocols and use clean materials to harvest mushrooms. This CCP was developed to minimize contamination with *S. aureus*, which is common on the skin of most individuals.

Transport. CCPs 4-M and 5-MC were developed to eliminate the potential for *S. aureus* outgrowth and protect the mushrooms from undesirable microbial growth and chemical contamination. Following grading and weighing at collection stations, the mushrooms are packed into cleaned and sanitized open boxes (resembling milk carton shippers) and/or perforated bags. The bags will not be sealed. This CCP, 4-M, must be followed. Placing mushrooms in sealed bags greatly increases the potential for toxin development (Hardt-English et al. 1990). CCP 5-MC sets a limit on shipping time. Mushrooms which exceed minimum periods will be rejected. Since trucks are not refrigerated, this must be adhered to.

Processing. The first CCP in the process flow is 6-MPC. This is an incoming inspection CCP designed to prevent potential problems from entering the plant. Product in sealed bags, which did not reach the plant within set times, abused product, and mushrooms which may have been contaminated with chemicals or hazardous physical materials will be rejected. The PRC, unlike the United States, does not believe in using materials once. All bags and boxes are reused, unless they are damaged, and, therefore, must be cleaned and sanitized, and tagged to indicate they have been used before being sent back for reuse (7-M). One hypothesis for a contributing cause for enterotoxin development was the reuse of insanitary bags, which inoculated fresh product with large numbers of viable *S. aureus*. The prevailing theory is that staphylococcal enterotoxin formed in product before it entered the plant, and was not removed or destroyed in the actual process flow. Strict adherence to these first seven CCPs should eliminate the enterotoxin concern.

CCP 8-M was instituted to ensure proper blanching. This CCP includes blanch time and temperature. No unauthorized changes shall be made to the system. Continuous temperature indicating devices are monitored and must be standardized at set intervals.

The next CCP, 9-MPC(S) is a general sanitation CCP and addresses microbiological, physical, and chemical issues. This CCP established set cleaning and maintenance protocols for the plant, equipment, flumes, and tanks, and general sanitation practices for all plant personnel. Tasks must be conducted according to set procedures and signed off by sanitation staff before processing can be initiated. This particular CCP may be called basic Good Manufacturing Practices, but FDA concerns and a need for continuing education in this area led to this being set as a distinct safety concern.

CCP 10-MC was put in place as an in-plant check. Damaged and/or poor quality products are removed from the process and destroyed. Poor mushroom quality could indicate that the product was abused in transit. CCP 11-M specifically addresses slicer maintenance. These units must be fully broken down and cleaned daily, and rinsed every two hours during production. Records of the cleanups and rinses must be retained.

With the exception of CCP 13-MPC, CCPs 12–20 were all put in place to ensure compliance with the FDA's low-acid canned food regulations (FDA 1989). CCP 13-MPC addresses operation of the can washers before filling and sealing. These washers must be turned on, time and temperature of operation monitored continuously, recorders must be monitored and maintained, and standardized as needed. The remaining CCPs in this block relate to fill weights (14-M), brining (15-M), can manufacture and seam integrity (12 and 17-M), coding (16-MPC), initial temperatures (18-M), retort process, retort configuration, and temperature indicating device calibration and maintenance (19-M). Those CCPs related to establishing processing parameters (fill weight, process time and temperature, etc.) must be determined by recognized process authorities. CCP 20-M pertains to chlorination of and residual chlorine levels in cooling water. Records are maintained for each CCP according to Food and Drug regulations.

Post-process. The final five CCPs were designed to protect the product and ensure that no suspect or unsafe product is shipped. CCP 21-M was initiated to protect the retorted cans from contamination after process. Equipment design and maintenance protocols have been established. Staff involved in handling processed cans have also been provided with basic background in container handling. CCP 22 and 23-M are microbiological monitoring critical control points. All processed product is incubated for a set time at 25°C. During incubation all cans are observed by trained staff for evidence of swelling. The discovery of swells results in the lot containing them being placed on hold. The next CCP entails 100% tap toning of the processed product by well-trained staff. The discovery of low vacuum containers also prompts a hold of the suspect lot.

CCP 24-MPC is, like CCP 16-MPC, a coding critical control point. Codes are essential for proper tracking and recovery of product. It is also required for low-acid canned foods.

The final point is container loading. All mushrooms exported from the PRC are shipped via container. The exporters must know what is in each container, and ensure that it is properly loaded to reduce the risk of damage in shipment.

Summary. This HACCP program for mushrooms produced in the PRC, specifically Fujian Province, was developed in response to the discovery of staphylococcal enterotoxin in product produced in that country. There are CCPs in this program which may evolve into simply control points for quality maintenance. The program presented here has been effective in controlling the staph concerns. The program does include record keeping requirements, criteria for detecting and evaluating deviations, and verification activities. The primary verification activity has been extensive for finished product sampling for the presence of toxin using the TECRA Elisa test.

Shredded lettuce

It was not so long ago that people purchasing lettuce, or other vegetables for that matter, for restaurants, the home or cafeterias, would buy those items directly from a produce dealer. Today, the added work and expense for these same restauranteurs or consumers to shred lettuce or cut carrots is considered wasteful. There are now many varieties and variations of pre-cut vegetables available to the consumer, but learning and applying the technology for developing and servicing this market has not been easy.

The produce grower and buyer have always known that the keys to meeting their market demands were to harvest, chill, and ship the product quickly. Expanding their operations to supply pre-cut products has required a rethinking of this traditional approach. The keys now are to keep their operations clean and the product cold. Losing control on either will compromise product quality and safety. Cutting vegetables releases cellular components, which can provide nutrients for microorganisms, and spread organisms over the cut surfaces and elsewhere.

A major concern with these pre-cut refrigerated products is the presence and growth of pathogenic organisms, particularly *Listeria monocytogenes*. Listeria has been isolated from lettuce (Sizmur and Walker 1988; Steinbrugge et al. 1988) and grows on cauliflower, broccoli, asparagus, (Berrang et al. 1989) and cabbage (Kallander et al. 1991). Shredded cabbage will also support the growth of other pathogens, such as *S. sonnei* (Satchell et al. 1990). There is also some concern related to hepatitis (lettuce). Since these products are not processed, it is essential to process and handle them properly to reduce the initial loads and to minimize the potential for recontamination and outgrowth of food pathogens.

The next HACCP model was developed for shredded lettuce packaged in gas permeable bags. The process was modeled on those used by California lettuce packers. Lettuce is a unique product in that it is extremely sensitive to abuse conditions, such as temperature, insanitation, or mishandling, and will show it. It is because of this characteristic that some individuals believe lettuce is not and should not be treated as a hazardous product. Product that looks and smells good can cause illness, however. It is for this reason and the fact that lettuce is not processed to destroy organisms that HACCP is a necessity.

Product description. The product may be described as follows: Iceberg lettuce is cored and shredded and packed into gas permeable bags continuously formed from roll stock. No modified atmospheres are used. The product is packaged for both the retail and institutional markets. The product is obtained from growers in the Salinas Valley in California, who are inspected by the company agronomists. Each bag and case is coded with a product code and "Use By" date. Individual bags and cases are marked "KEEP REFRIGERATED/ STORE AT LESS THAN 40°F".

Risk assessment. *Microbiological.* Since there is no kill step or application of inhibitory chemicals or gases which are used in many processes to reduce the numbers of or destroy microorganisms, any organisms, pathogenic or not, which are present on the product initially or which contaminate the product during processing or handling, will remain on the product (see Fig. 14-5). The primary organisms of concern are *Salmonella* sp. and *Listeria monocytogenes*, the latter having been identified as a common contaminant in many food plants. The keys to this process are to keep the product cold and keep the processing lines and plants clean.

Chemical and physical. The chemical and physical risk assessments for the bagged and fresh lettuce may be seen in Figs. 14-6 and 14-7. Major concerns in these two areas are pesticides and metal contamination.

Product	Microbiological Hazard Characteristics Associated with the Food (+ for Yes; 0 for No)						Hazard Category
	A High Risk Special Population	B Sensitive Ingredients	C No Kill-Step in Process	D Recontam. between Proc/Pack	E Abusive Handling Dist/Cons	F No Term. Heat Proc. by Consumer	
Shredded Lettuce	0	+	+	+	+	+	V
Raw Materials and Ingredients...As Received, before any Manufacturing Steps by the Food Facility (such as cooking)...							
Raw Material or Ingredient	A	B	C	D	E	F: No Kill step before receipt*	Hazard Category
Lettuce	0	+	+	+	+	+	V

*No heat process or any other kill-step applied after packaging by supplier;
No heat process or other kill-step before entering food plant.

FIGURE 14-5.   Assessment of microbiological risk for shredded lettuce.

Food Item	Hazard Characteristics Known to be Associated with the Food and It's Ingredients (+ for Yes; 0 for No)						Hazard Category
	A High Risk Special Population	B Ingredients Contain Hazard	C Not Re-moved in Manufact.	D Recontam. between Mfg/Pack.	E Contam. by Dist. or Cons.	F Cons. can-not Detect/ Rem.	
(1) Product							
Shredded Lettuce	0	+	+	+	0	+	IV
(2) Raw Mat's and Ing's							
Lettuce	0	+	+	+	+	+	V

Notes: (1) As used by consumer: (2) As entering the food facility before preparation or processing.

FIGURE 14-6.   Assessment of chemical risk for shredded lettuce.

Critical control points and limits. The process flow chart with Critical Control Points may be seen in Fig. 14-8. Like the mushroom model, development and implementation of CCPs begins all the way back in the field. The CCPs will be divided into three groups: Field, Process and Packaging, and Post-Packaging. As mentioned earlier, lettuce is a product to which no kill step is applied. Control of microorganisms is dependent on initial load, maintenance of good sanitation practices, and temperature.

Food Item	Hazard Characteristics Known to be Associated with the Food and It's Ingredients (+ for Yes; 0 for No)						Hazard Category
	A High Risk Special Population	B Ingredients Contain Hazard	C Not Re-moved in Manufac.	D Recontam. between Mfg/Pack	E Contam. by Dist. or Cons.	F Cons. Can-not De-tect/Rem.	
(1) Product							
Shredded Lettuce	0	+	0	+	0	+	III
(2) Raw Mat's and Ing's							
Lettuce	0	+	+	+	+	+	V

Notes: (1) As used by consumer; (2) As entering the food facility before preparation or processing.

FIGURE 14-7.   Assessment of physical risk of shredded lettuce.

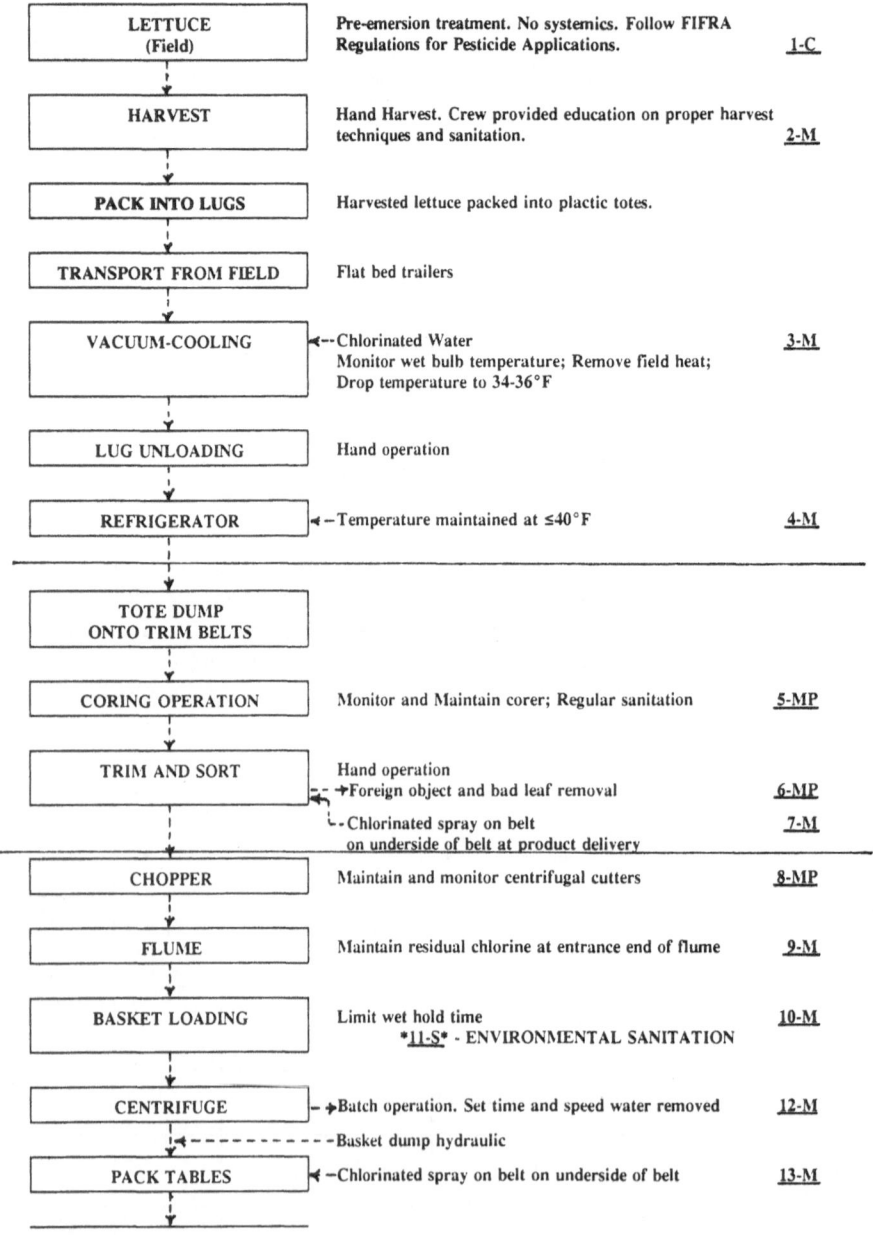

FIGURE 14-8.   Production of shredded lettuce. (M = Microbiological; C = Chemical; P = Physical; S = Sanitation)

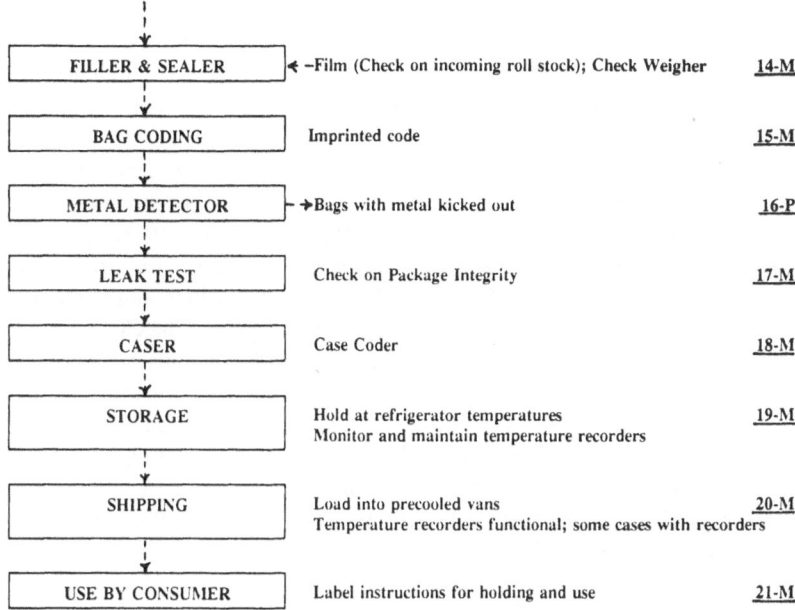

FILLER & SEALER	◄ –Film (Check on incoming roll stock); Check Weigher	<u>14-M</u>
BAG CODING	Imprinted code	<u>15-M</u>
METAL DETECTOR	┣ ➔Bags with metal kicked out	<u>16-P</u>
LEAK TEST	Check on Package Integrity	<u>17-M</u>
CASER	Case Coder	<u>18-M</u>
STORAGE	Hold at refrigerator temperatures Monitor and maintain temperature recorders	<u>19-M</u>
SHIPPING	Load into precooled vans Temperature recorders functional; some cases with recorders	<u>20-M</u>
USE BY CONSUMER	Label instructions for holding and use	<u>21-M</u>

FIGURE 14-8.   Continued.

Field. The first CCP, 1-C, is a control point for potential chemical hazards. Pesticides must be applied according to FIFRA regulations, and there must be documentation to confirm this. In fact, farmers must maintain accurate records of the use of all agricultural chemicals, including fertilizers.

CCP 2-M has been implemented in an effort to reduce initial loads on the product. The crew doing the harvest are provided with a basic education in proper harvest techniques, including routine sanitation of harvest tools. The grower must also provide adequate sanitary facilities for his field crew. This education provides an added benefit in that product quality is enhanced. This CCP is one that cannot be continuously monitored, but emphasis on proper employee education can contribute to its being self monitoring.

The next two CCPs are designed to reduce the field heat of the lettuce to 34–36°F and to maintain that temperature. Temperature is easily monitored, as is chlorination of cooling water.

Process and packaging. CCPs 5-MP to 15-M make up the process and packaging CCPs. Lettuce is a product, which is extremely sensitive to abuse conditions. The controls to ensure safety will also help to maintain quality. There are three CCPs in this section, 5-MP, 6-MP, and 8-MP, which have been initiated to control potentially harmful microorganisms and physical hazards. The keys to the former are proper equipment design and cleaning, which includes rinsing during operations. For 5 and 8, control of physical hazards is a maintenance issue. Cutters must be kept sharp, and replaced if broken. If breaks occur, the metal must be recovered.

CCPs 7-M, 9-M, and 13-M all relate to maintaining chlorine levels in water systems. These are monitored continuously and adjusted as needed.

The CCPs for the basket loading and centrifuge operations, 10-M and 12-M are timed. If the established times are exceeded, the product is discarded.

CCP 11-S is a general sanitation CCP. This is a CCP which some consider a good manufacturing practice and not a critical control point. The CCP includes education of cleaning crews and maintenance staff, adherence to cleaning protocols, and a dicta that the operation cannot be started until established cleaning and sanitizing protocols have been completed and reviewed by management.

The final two CCPs 14-M and 15-MPC, are related to packaging. Incoming rollstock is evaluated as it comes in and not approved for use if it fails to meet set criteria. 15-MPC is a coding CCP. Any product which is not coded will be rejected, as traceability will be lost.

Post-package. The final CCPs are designed to protect the packaged product, to help assure that no potentially unsafe product leaves the plant, and to ensure traceability. CCPs 16-M and 17-P, are aimed at assuring product safety. 16-P has been implemented to assure proper operation of the on-line metal detector. Each package is scanned and those with metal are removed. Parts of the CCP for this operation include maintaining and standardizing the unit. 17-M is a similar CCP. Each package is checked for integrity, that is, good seals to prevent contamination of the bagged lettuce.

The rationale for case coding (18-MPC) is the same as that for package coding; traceability.

CCPs 19 and 20-M have been developed to ensure that temperatures of refrigerators and vans are maintained and monitored. These CCPs include criteria for maintaining and standardizing temperature recorders, and ensuring that these maintenance records are kept. The final CCP is designed to protect the consumer, whether that is an individual or a restaurant. Storage and handling instructions are on all cases and individual containers. This, again, is not something which can be easily monitored, but it must be done to protect the consumer and the packer. Packers can place time/temperature indicators on packages. The indicator will change if the product is temperature abused, thus providing a measure of safety for the consumer.

Summary. This model program was developed after consultation with lettuce growers in the Salinas Valley in California. Such a model has not, as yet, been implemented, but it is under consideration. The flow chart includes unit operations, which may not be used in all operations, but are considered by technical staff to be essential for product safety. There are also programs, such as CCP 2-M, which have not been and may never be implemented because of labor issues, but are considered vital to product safety. It has already been noted that lettuce is a very sensitive product. Abuse frequently leads to deterioration of product quality, and loss of product. Using the HACCP approach to assure product safety will help to enhance product quality. With shredded lettuce, the two go hand-in-hand.

Frozen French fries (crinkle cut)

The French fry is a common accompaniment to meals the world over. They may be produced for consumption directly from fresh potatoes, from partially fried frozen potato strips, or from frozen formulated strips. French fries are by far the largest selling item at fast food restaurants, which are the largest consumer of the frozen par-fried products. As an example, in 1987 the frozen potato pack in the United States was 5,287,634 thousand pounds of which 4,539,795 thousand pounds were French fries. Approximately 87% (3,951,099 thousand pounds) were produced for sale to the foodservice industry (Anonymous 1988).

This model will focus on French fries or par-fries produced for the foodservice and retail markets from whole potatoes, not the formulated products. This simple product is defined in the United States Standards, which includes styles of cut, strip length, and grading and preparation parameters (Anonymous 1988). The simplicity of the product, the industrial practices used to manufacture it, and how French fries are prepared for serving, all contribute to making this a relatively safe product. This does not imply that things cannot go wrong. Things do, so therefore, HACCP systems for this particular product are being or have been put in place throughout the country. Remember, the key with HACCP is prevention. It is a proactive program.

The process flow chart and critical control points were developed from published literature (Blackstock and Skiver 1974) and from discussions with individuals in the industry. This model will focus on product manufactured for the institutional market, which is defined as packages of five pounds or more, although the process flow is the same regardless of ultimate destination.

Product description. Since this is a product which has an established U.S. grade standard, the product description will mirror that described in the standards.

Frozen crinkle-cut French fried potatoes are prepared from mature, sound white or Irish potatoes. The potatoes are washed, sorted and trimmed as necessary to assure a wholesome product. The strips are partially fried in a continuous fryer in vegetable oil (soybean, canola, or palm olein). Before frying, the potatoes are dragged through a tank containing dextrose and dipotassium phosphate solution. The partially fried products are packed for the food service market into

5-pound polyethylene lined brown paper bags, which are glued closed and packed six to a case.

The product has a one year frozen shelf life. Both individual bags and cases are coded with a "Use-By" date, lot code, and ingredients statement (potatoes, oil, dextrose, and dipotassium phosphate). Cases bear storage, handling, and preparation instructions. The French fries may be deep-fried in hot oil (365°F) for 1 1/2 minutes or until the larger pieces are fully cooked; or baked in an oven, which has been pre-heated to 425°F for 15 minutes. For oven baking, the fries are spread one layer think (Anonymous 1988).

Risk assessment. *Microbiological.* The risk assessment for microbiological concerns may be seen in Fig. 14-9. Frozen French fried potatoes are not what would be considered a hazardous product. The deep frying process, usually at 375°F or above, will destroy vegetative cells of pathogens and non-pathogens. Before serving, the product is finished fried or baked, which will again inactivate most microorganisms.

With the exception of the potatoes, the ingredients are not considered hazardous either. Potatoes are grown in the ground, and are, therefore, filthy. Potatoes would pose a greater concern if they were being canned.

Product	Microbiological Hazard Characteristics Associated with the Food (+ for Yes; 0 for No)						Hazard Category
	A High Risk Special Population	B Sensitive Ingredients	C No Kill-Step in Process	D Recontam. between Proc/Pack.	E Abusive Handling Dist/Cons.	F No Term. Heat Proc. by Consumer	
French Fries	0	+	0	+	+/0	0	II/III
Raw Materials and Ingredients...As received, before any manufacturing Steps by the Food Facility (such as cooking)...							
Raw Material or Ingredient	A	B	C	D	E	F: No Kill Step Before Receipt*	Hazard Category
Potatoes	0	+	+	+	+	+	V
Oil	0	0	0	+	0	0	I
Dextrose	0	0	0	0	0	0	0
Dipotassium Phosphate	0	0	0	0	0	0	0

*No heat process or any other kill-step applied after packaging by supplier;
No heat process or other kill-step before entering food plant.

FIGURE 14-9.   Assessment of microbiological risk of French fries.

Food Item	Hazard Characteristics known to be Associated with the Food and It's Ingredients (+ for Yes; 0 for No)						Hazard Category
(1) Product	A High Risk Special Population	B Ingredients Contain Hazard	C Not Re- moved in Manufact.	D Recontam. between Mfg/Pack.	E Contam. by Dist. or Cons.	F Cons. Can- not De- teck/Rem.	
French Fries	0	+	+	+ /0	0	+	III/IV
(2) Raw Mat's and Ing's							
Potatoes	0	+	+	+	+	+	V
Oil	0	0	0	+	+ /0	+	II/III
Dextrose	0	0	0	0	0	0	0
Dipotassium Phosphate	0	0	0	0	0	0	0

Notes: (1) as used by consumer; (2) as entering the food facility before preparation or processing.

FIGURE 14-10.   Assessment of chemical risk of French fries.

Chemical and physical. The physical and chemical risk assessments for the French fries may be seen in Fig. 14-10 and 14-11. The major chemical concern would be pesticides. Producers are also concerned about metal fragments so most lines include metal detectors.

Critical control points and limits. The process flow with the associated Critical Control Points may be seen in Fig. 14-12. Of the three models, this is by far

Food Item	Hazard Characteristics Known to be Associated with the Food and It's Ingredients (+ for Yes; 0 for No)						Hazard Category
(1) Product	A High Risk Special Population	B Ingredients Contain Hazard	C Not Re- moved in Manufact.	D Recontam. between Mfg/Pack.	E Contam. by Dist. or Cons.	F Cons. Can- not De- tect/Rem.	
French Fries	0	+	0	+ /0	0	+	II/III
(2) Raw Mat's and Ing's							
Potatoes	0	+	+	+	+	+ /0	IV/V
Oil	0	0	0	0	0	0	0
Dextrose	0	0	0	0	0	0	0
Dipotassium Phosphate	0	0	0	0	0	0	0

Notes: (1) As used by consumer; (2) As entering the food facility before preparation or processing.

FIGURE 14-11.   Assessment of physical risk of French fries.

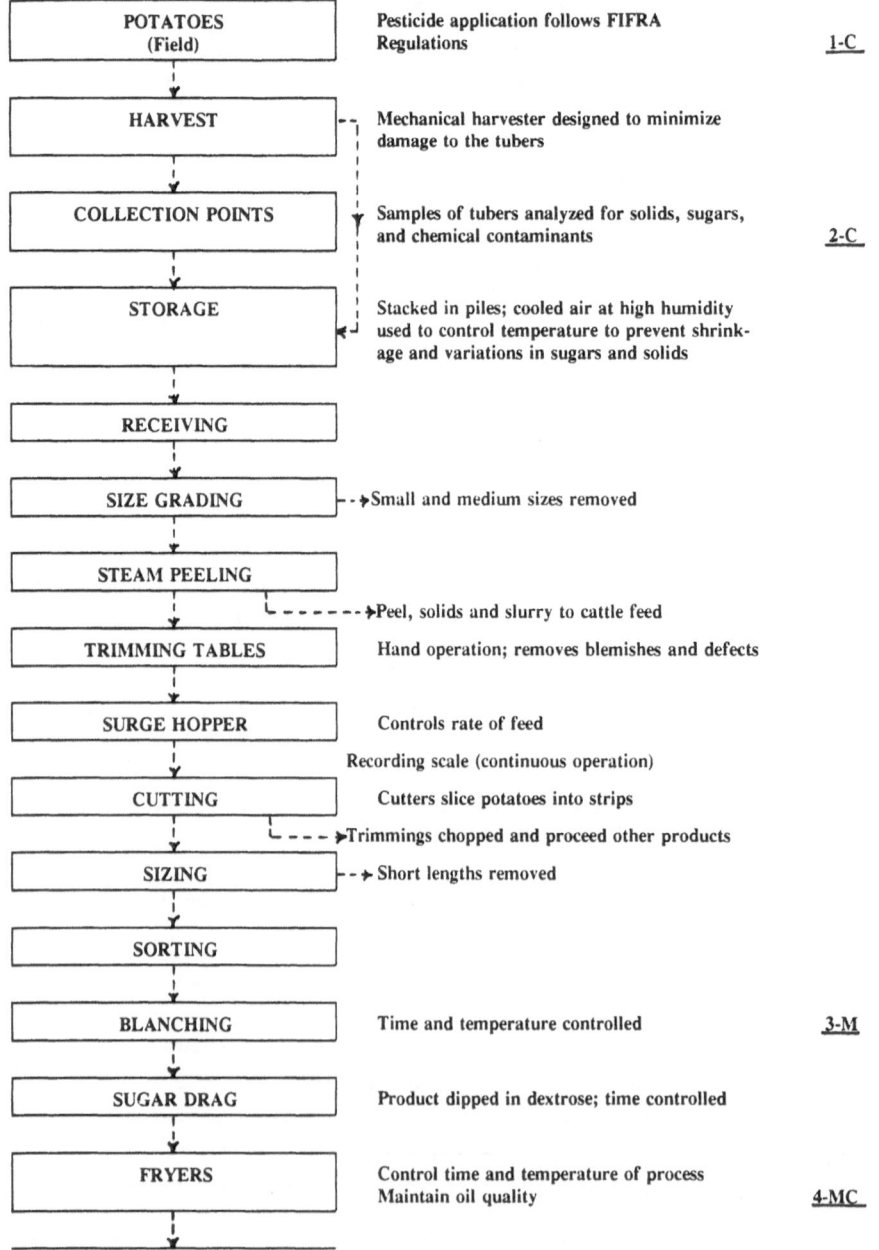

FIGURE 14-12.   French fry processing. (M = Microbiological; C = Chemical; P = Physical; S = Sanitation)

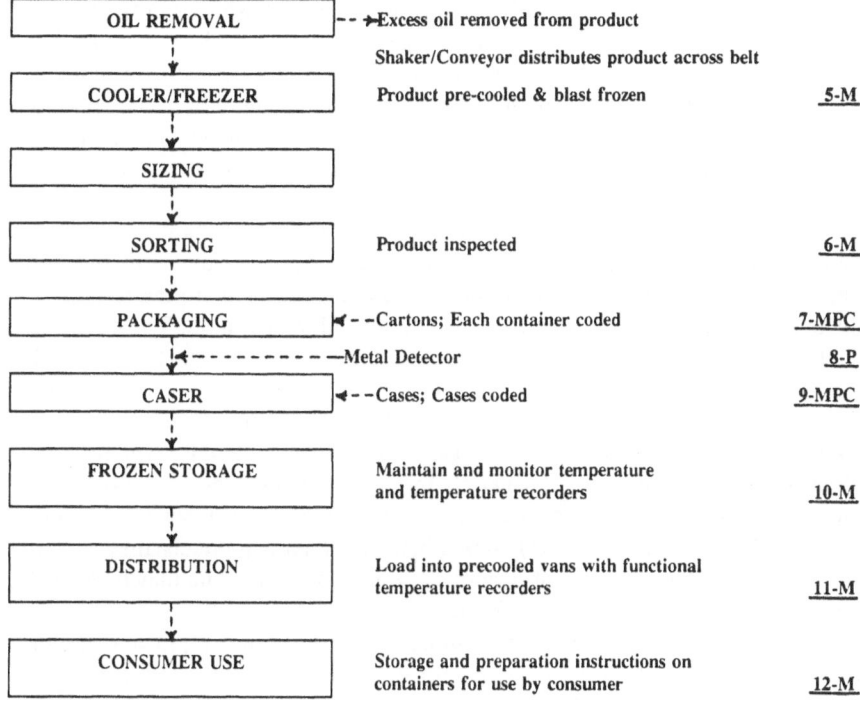

FIGURE 14-12.   Continued.

the simplest with only twelve CCPs. The twelve points will be divided into three groups: Cultivation and Storage, Processing and Packaging, and Post-Process Handling.

Cultivation and storage. The first CCP, 1-C, has been implemented to monitor pesticides and other agricultural chemicals. All growers supplying potatoes to this operation must maintain records of pesticide and fertilizer applications, records which are regularly reviewed by company agronomists. Applications of pesticides must also adhere to FIFRA regulations. The crops of growers whose records are incomplete or non-existent will not be purchased.

Unlike the products used in the previous models, lettuce and mushrooms, raw potatoes are very hardy. They can be held for long periods before being processed or eaten. This characteristic is utilized in CCP 2-C. While in storage, lot samples are tested for both quality parameters, which affect finished product acceptability (not part of the HACCP plan) and pesticide levels. The latter is the CCP. Storage conditions are also strictly monitored to control quality parameters.

Processing and packaging. The first process step for which a critical control point has been established is the blanching operation, CCP 3-M. Temperatures are monitored continuously at two points on each line, and automatically adjusted; belt speeds are controlled and checked routinely and cannot be changed during operation; and temperature recording devices are monitored and standardized on a regular schedule by an outside process authority. Records are maintained for all standardization protocols.

CCP 4-MC is a control point for the deep fat frying operation. Blanched potatoes are fried at temperatures which will destroy vegetative cells of micro-organisms. The factors which are monitored and controlled are the same as for blanching, that is, frying time (belt speed) and oil temperature. The controllers and recorders are also monitored and standardized routinely. There are also controls for certain chemical parameters of the oil. The oil is tested before frying is initiated to ensure that the alkali cleaners have been removed from the system. Degradation products of frying are also tested. Polar materials, which may be defined as all nontriglyceride materials in the oil, are considered the best index of oil quality and are, in fact, used by regulatory agencies in several European nations (Firestone et al. 1991), but other degradation products may be used (Blumenthal 1987). The use of severely degraded oils for frying may have health implications (Clark and Serbia 1991).

The next CCP, 5-M, has been implemented to ensure rapid and uniform cooling (freezing) of the par-fried potato strips. Belt speed and freezer temperatures are maintained and recorded continuously. Recorders are checked and standardized regularly, and records of this operation maintained.

CCP 6-M is in place in an effort to control the only hand operation in the line. To minimize recontamination at the sorting table, all employees are required to wear sanitized gloves and maintain sanitary conditions. The product receives no further processing by the manufacturer from this point on (it *should* be fried or baked by the user), so contamination is a concern.

The final CCP in the processing and packaging area is CCP 7-MPC. This is the control point for coding. Each package of French fries is given a unique product and period code and is marked with a "Use-By" date. Uncoded units are set aside for rework, destruction, or manual coding. Adherence to strict coding protocols enhances traceability should any problems arise protecting the consumer and the manufacturer.

Post-process handling. The final five CCPs have been implemented to protect the packaged product during storage, distribution, and preparation by the consumer. The first, CCP 8-P, has been put in place to reduce the risk of injury due to metal contamination. Each package is passed through a metal detector, which removes those with metal fragments. The metal detector must be turned on (easily visible green and red lights signifying on and off) and it is tested regularly using standards with known levels and sizes of contaminants. Test results and maintenance records are retained.

CCP 9-MPC is a critical control point for case coding. The same criteria as described for 7-MPC apply here. Being an institutional product, the case code is more important than the package code. Unlike cases containing retail products, the individual packages in an institutional product will be used in one location.

CCP 10 and 11-M are designed to ensure that the product does not suffer from temperature abuse. In both the cold storage warehouse and the vans, working temperature recorders are required. These units must be maintained and standardized regularly. With a frozen product, the key is to prevent the product from thawing and warming to a point where outgrowth of pathogens will occur. Refreezing the now contaminated product can be hazardous. *S. aureus* has the ability to grow and produce toxin in frozen foods which have been abused. This toxin may not be destroyed in the cooking process.

The final critical control point is the preparation, 12-M. These are designed to protect both the consumer and the manufacturer. Instructions should be clearly written so that they may be easily understood by the user. Since French fries are not supposed to be eaten raw and should be fully cooked before consumption, microbiological concerns are minimal, but they are there and must be addressed.

Summary. French fries have been the simplest model presented. It is also the safest product. This is a product which requires a HACCP plan, however. All one needs to do is look at the volumes of French fries produced and consumed to understand why.

Refrigerated chicken salad

Chicken salad is a product which is frequently implicated as a vehicle in foodborne illness (Bryan 1988a). There are numerous reasons for this, several of which Bryan (1988b) included in "top ten" causes for foodborne illness. Causes of outbreaks from chicken salad include but are not limited to improper refrigeration, recontamination of processed product, handling of processed product by infected people, use of contaminated raw materials, and mishandling by consumers or retailers, all of which are violations of basic food safety principles. Among the pathogenic microorganisms which have been associated with chicken are salmonella, *S. aureus,* and *Listeria monocytogenes.* The latter organism is of greatest concern at this time because of its ability to grow at low temperatures, survival at reduced pH levels, and its recognition as an environmental contaminant of food plants, especially with drains, coolers, and air conditioners.

The potential problems inherent in prepared refrigerated foods, such as chicken salad or sous vide items, has been acknowledged by both the industry and regulatory agencies. The Food & Drug Administration has, in fact, instituted a series of workshops on vacuum-packaged foods, which have and will address some of these concerns. Trade associations have helped spearhead industry response in this area. For example, the Salad Manufacturers Association has offered HACCP courses to their members using real world models as teaching aides. One model was chicken salad (Corlett and Stier 1991b). The model used in the class served two purposes: it was first used as a class exercise and then served as the basis for industry attendees to develop a generic HACCP model. With their input, the generic model incorporated a wide range of ingredients and industry practices. This model was prepared by Corlett and Mitchell (1991) on behalf of the Salad Manufacturers Association, and has been presented to the USDA/FSIS. The model serves as the next example. The model includes times and temperatures; sets a maximum product pH of 5.5; sets maximum storage temperatures at 45°F; and does not advocate the use of controlled or modified atmosphere packaging. This model is much more complex than the previous three. It provides an excellent example of the complexities involved in developing a HACCP plan for complex, formulated products. This section follows the HACCP plan submitted by Corlett and Mitchell (1991) very closely and is included in this chapter with their permission.

Introduction. When developing this HACCP model, the salad manufacturers who participated discussed a wide variety of issues related to production of chicken salad; therefore, the final generic model included the following, written to help clarify the model (Corlett and Mitchell 1991). These are quoted verbatim.

1. There are fifteen critical control points that were expanded into 39 critical limits that must be monitored and are subject to deviation control and record keeping. In certain instances, some may be eliminated from the list if the identified hazard is eliminated by safe design of the system.

2. Critical control point number 4 applies to processing treatments to *reduce* the contamination on raw materials such as celery and onions. Treatments may include controlled washing with chlorinated water or blanching and are intended to be a kill step. Controlled processing treatments are appropriate for reducing the risk of microbiological contamination on raw agricultural products used in ready-to-eat products.

3. Each food plant or company should have a "Food Safety Deviation Authority" that determines the disposition of product placed on hold when monitoring indicated that a critical control point or critical limit was out of control. This is essential to assess the safe disposition of the held food and to use judgment

in determining whether the lot is subject to further testing, may be released, or
destroyed.

4. Open date product shelf-life coding is required. It has not been resolved
whether a "sell-by" or a "use-by" date is to be used.

5. Minimum requirements for sanitation and food handling must be based on
USDA/FSIS procedures. This will provide uniformity for all users of the model.

Product description. The following product description for chicken salad was
developed (Corlett and Mitchell 1991):

> Refrigerated chicken salad is intended for and marketed in retail grocery stores
> and in foodservice establishments. It is a ready-to-eat product prepared from a
> combination of fresh and processed ingredients. Product is packaged in similar

Raw Material or Ingredient	How Received or Prepared	Raw Material or Ingredient	How Received or Prepared
Cooked chicken (from USDA Establishment)	•Frozen, or •Canned	Starch	•Flour
Dressing	•Ready-made, or, •Prepared in manufacturing facility	Gums and stabilizers	•Dehydrated
Celery (diced)	•Fresh stalk celery in crates, or •Frozen, or •Canned	Lemon juice	•Concentrated
Bread Crumbs/Cracker Meal	•Ready-to-use	Horseradish	•Prepared/acidified
Diced sweet pickles and pickle relish	•Ready-to-use	High fructose corn syrup	•Liquid
Red or green bell peppers	•Fresh, or •Frozen, or •Dehydrated, or •Canned	MSG/HVP	•Dry powder
Hard boiled eggs (diced)	•Purchased hard boiled and peeled	Salt	•Crystalline
Diced onion	•Fresh, or •Frozen, or •Dehydrated •Concentrated	Sugar	•Crystalline
Chicken broth	•Concentrated	Citric acid	•Crystalline
Onion powder	•Dehydrated	Titanium dioxide	•Powder
Garlic powder	•Dehydrated	Textured vegetable protein	•Frozen, or •Dehydrated
Spices	•Dehydrated	Natural and artificial flavors	•Powder

FIGURE 14-13.   Refrigerated chicken salad ingredients.

consumer or foodservice containers, usually consisting of tubs with snap-on lids or plastic bags. No control or modified atmosphere packaging (CAP/MAP) is applied during packaging. Shelf-life is controlled by refrigeration at 45°F or less with appropriate open-date coding. The label or containers include this code, as well as the words "Keep Refrigerated." Distribution and retail storage temperature of 45°F and designated shelf life limit, must not be exceeded (Corlett and Mitchell 1991).

This is a complex product, which uses many different ingredients. The ingredients list included in the model are itemized in Fig. 14-13.

Risk assessment. It is essential that risk on both the product and the ingredients be determined. With so many ingredients, this exercise becomes more difficult. Figures 14-14, 14-15, and 14-16 show the risk assessment information developed by the salad manufacturers and submitted by Corlett and Mitchell (1991).

Critical control points and limits. The critical control points for this model may be seen on the process flow chart for refrigerated chicken salad, Fig. 14-17. In the three previous models, the CCPs and limits were discussed. In this model, the limits for each critical control point are described in Fig. 14-18. The format used in this figure, which includes the CCP number, a description of each CCP, and the limits, was developed by Corlett and Stier (1991b) and has been used extensively as a teaching aide. When the time comes for developing a HACCP plan, this kind of format allows users to see how the plan is set up. It is also organized and allows for review, both by internal staff and the agencies.

Summary. This particular chicken salad model is even more extensive than presented. The developers established criteria for monitoring the critical control points, evaluating deviations, record keeping, and verifying that the HACCP system is actually working. It is a model for a complete HACCP plan and is, in fact, being used in the industry. These exercises were assembled to give processors ideas on how to "get going" on the first three principles, that is, risk assessment, establishing CCPs, and setting limits. The final four principles will therefore not be addressed.

FIGURE 14-14.   Assessment of microbiological risk in refrigerated chicken salad.

* No heat process or any other kill-step applied after packaging by supplier;
  No heat process or other kill-step before entering food plant.
ESCAgenetics Corporation, Food Safety Division, D.A. Corlett, 01/14/91. FT/DOC.HACCPMIC

Product	Microbiological Hazard Characteristics Associated with the Food (+ for Yes; 0 for No)						Hazard Category
	A High Risk Special Population	B Sensitive Ingredients	C No Kill-Step in Process	D Recontam. between Proc/Pack.	E Abusive Handling Dist/Cons	F No Term. Heat Process by Consumer	
Salad	0	+	+	+	+	+	V

Raw Materials and Ingredients...as Received, Before any Manufacturing Steps by the Food Facility (such as cooking)							
Raw Mat. or Ingred.	A	B	C	D	E	F: No kill Step Before Receipt*	Hazard Category
Cooked Chicken	0 0	+ +	0 Frozen 0 Canned	+ 0	+ 0	0 0	III I
Dressing	0 0	+ Eggs 0 No Eggs	0 0	+ +	+ +	0 0	III II
Celery	0 0 0	+ + +	+ Fresh + Frozen 0 Canned	+ + 0	+ 0 0	+ + 0	V IV I
Bread Crumbs	0	0	0	+	0	0	I
Pickle R.	0	0	0	+	0	0	I
Red & Gr. Bell Peppers	0 0 0 0	+ + + +	+ Fresh + Frozen + Dehyd. 0 Canned	+ + + 0	+ 0 0 0	+ + + 0	V IV IV I
Eggs HBP	0	+	0	+	+	0	III
Onion, Diced	0 0 0	+ + +	+ Fresh + Frozen + Dehydr.	+ + +	+ 0 0	+ + +	V IV IV
Ch. Broth	0	+	0	+	0	0	II
Onion Pwdr	0	+	+	+	0	+	IV
Garlic Pwdr	0	+	+	+	0	+	IV
Spices	0 0	+ +	+ Natural 0 Treated	+ +	0 0	+ 0	IV II
Starch	0	0	0	+	0	0	I
Gums	0	0	+/0	=	0	+/0	III
Lemon J.	0	0	0	0	0	0	0
Horse Rad.	0	0	0	0	0	0	0
HF Corn S.	0	0	0	0	0	0	0
MSG	0	0	0	0	0	0	0
HVP	0	0	0	+	0	0	I
Salt	0	0	0	0	0	0	0
Sugar	0	0	0	0	0	0	0
Citric Acid	0	0	0	0	0	0	0
Tit. Diox.	0	0	0	0	0	0	0
Water	0 0	+ +	+ Not treat. 0 Treated	+ +	0 0	+ 0	IV II
Text. Veg. Protein	0 0	+ +	+ Frozen + Dehyd.	+ +	+ +	+ +	V V

(1) Product	Hazard Characteristics Known to be Associated with the Food and It's Ingredients (+ for Yes; 0 for No)						
	A High Risk Special Population	B Ingredients Contain Hazard	C Not Removed in Manufact.	D Recontam. between Mfg/Pack.	E Contam. by Dist. or Cons.	F Cons. Cannot Detect/Rem.	
Salad	0	+	+	0	+	+	IV
(2) Raw Mat's and Ing's							
Cooked Chicken	0	+	0 Frozen	0	+	+	III
	0	+	0 Canned	0	0	+	II
Dressing	0	+	+ Scratch	+	0	+	IV
	0	+	0 Prepared	0	0	+	II
Celery	0	+	+ Fresh	+	+	+	II
	0	+	0 Frozen	0	0	+	II
Bread Crumbs	0	+	+	0	0	+	III
Pickle Rel.	0	+	0	0	0	+	II
Red & Gr. Bell Peppers	0	+	+ Fresh	+	+	+	V
	0	+	0 Frozen	0	0	0	II
	0	+	0 Canned	0	0	0	II
	0	+	0 Dehydr.	0	0	0	II
Eggs HBP	0	+	+	+	+ /0	+	V/IV
Onion	0	+	+ Fresh	+	+	+	V
	0	+	0 Frozen	+	+	+	III
	0	+	0 Dehydr.	0	+	+	III
Chicken Broth	0	+	+	0	0	+	III
Onion Pwdr	0	+	0	0	0	+	II
Garlic Pwd	0	+	0	0	0	+	II
Spices	0	+	+ /0	+	+	+	V/IV
Starch	0	+	0	0	0	+	II
Gums	0	0	0	0	0	0	0
Lemon J.	0	+	0	0	0	0	I
Horse Rad.	0	+	0	0	0	+	II
HF Corn S	0	0	0	0	0	0	0
MSG	0	0	0	0	0	0	0
HVP	0	0	0	0	0	0	0
Salt	0	0	0	0	0	0	0
Sugar	0	0	0	0	0	0	0
Cit. Acid	0	0	0	0	0	0	0
Tit. Diox.	0	0	0	0	0	0	0
Water	0	+	+ Not treat.	0	+	+	IV
	0	+	0 Treated	0	+	+	III
Text. Veg. Protein	0	+	0	0	+	+	III
Flavors	0	0	0	0	0	0	0

Notes: (1) As used by consumer; (2) As entering the food facility before preparation or processing.

FIGURE 14-15. Assessment of chemical risk in refrigerated chicken salad.

Food Item	Hazard Characteristics Known to be Associated with the Food and It's Ingredients (+ for Yes; 0 for No)						Hazard Category
(1) Product	A High Risk Special Population	B Ingredients Contain Hazard	C Not Re- moved in Manufact.	D Recontam. between Mfg/Pack	E Contam. by Dist. or Cons.	F Cons. Can- not De- tect/Rem.	
Salad	0	+	0	+	+	+	IV
(2) Raw Mat's and Ing's							
Cooked Chicken	0 0	+ +	0 Frozen 0 Canned	0 0	0 0	+ +	II II
Dressing	0 0	+ +	+ Scratch 0 Prepared	0 0	0 0	= =	III II
Celery	0 0 0	+ + +	+ Fresh 0 Frozen 0 Canned	+ 0 0	+ + 0	+ + +	V III II
Bread Crumbs	0	+	0	+	0	+	III
Pickle Rel.	0	+	0	+	0	+	III
Red & Gr Bell Peppers	0 0 0 0	+ + + +	+ Fresh 0 Frozen 0 Dehyd. 0 Canned	+ 0 + 0	+ + + 0	+ + + +	V III IV II
Eggs HBP	0	+	0	0	+/0	0	II/I
Onion	0 0 0	+ + +	+ Fresh 0 Frozen 0 Dehyd.	+ 0 +	+ 0 0	+ + +	V II III
Chicken Br.	0	0	0	0	0	0	0
Onion Pwdr	0	+	0	+	0	+	III
Garlic Pwdr	0	+	0	+	0	+	III
Spices	0	+	+/0	+	0	+	IV/III
Starch	0	+	0	+	0	+	III
Gums	0	+	0	+	0	+	III
Lemon J.	0	0	0	0	0	0	0
Water	0	0	0	0	0	0	0
Horse Rad.	0	+	0	+	0	+	III
Corn St.	0	+	0	+	0	+	III
MSG	0	+	0	+	0	+	III
HVP	0	+	0	+	0	+	III
Salt	0	+	0	+	+	+	IV
Sugar	0	+	)	+	+	+	IV
Cit. Acid	0	+	0	+	0	+	III
Tit. Diox.	0	+	0	+	0	+	III
Text. Veg. Protein	0 0	+ +	0 Frozen 0 Dehydr.	+ +	0 0	+ +	III III
Flavors	0 0	+ +	0 Liquid 0 Powder	0 +	0 0	+ +	II III

Notes: (1) As used by consumer; (2) As entering the food facility before preparation or processing.

FIGURE 14-16. Assessment of physical risk in refrigerated chicken salad.

155

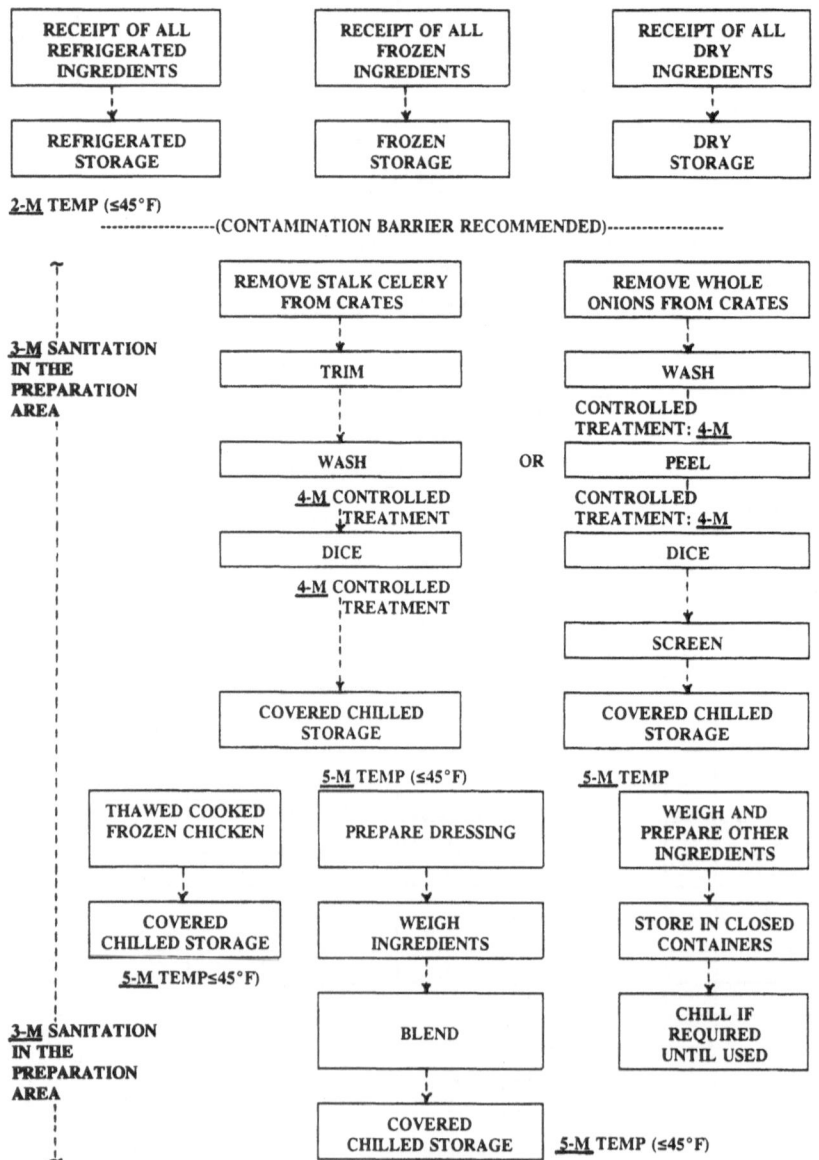

FIGURE 14-17. Production of refrigerated chicken salad. (M = Microbiological; C = Chemical; P = Physical; S = Sanitation)

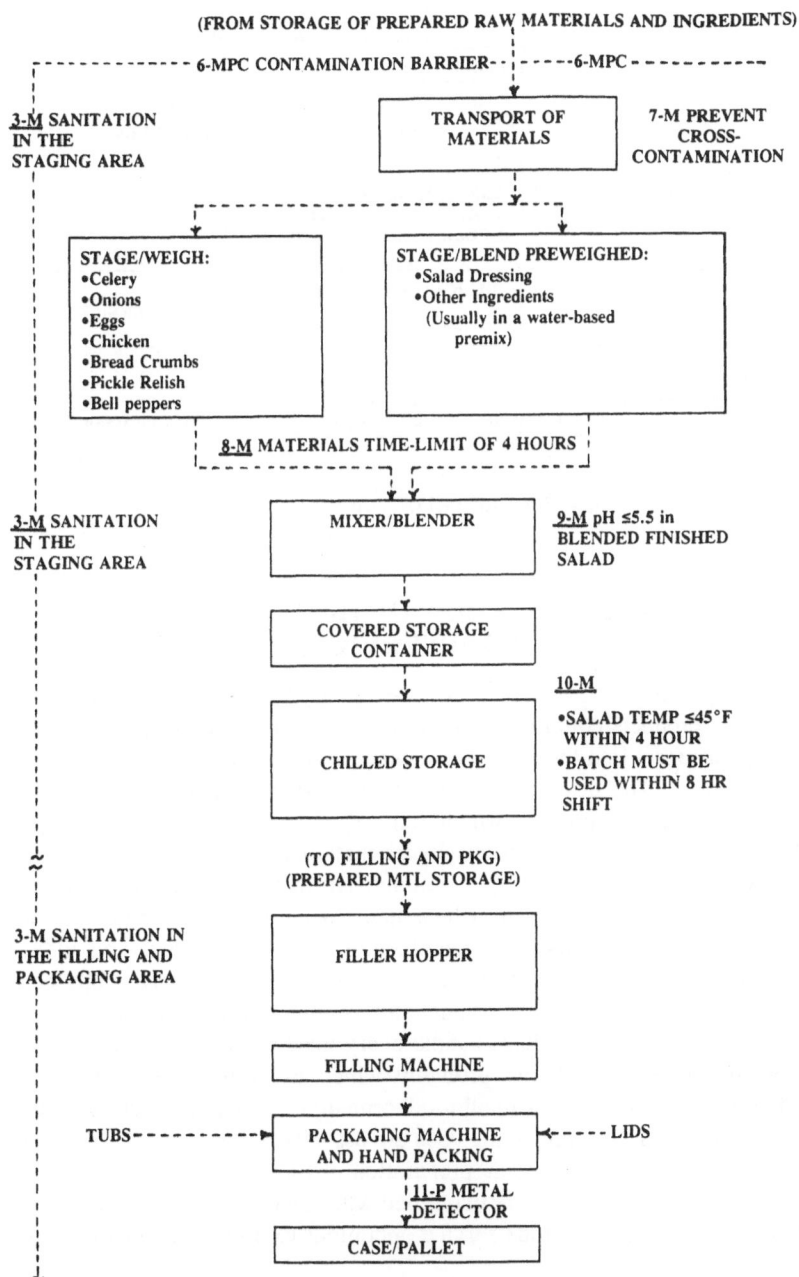

FIGURE 14.17. Continued.

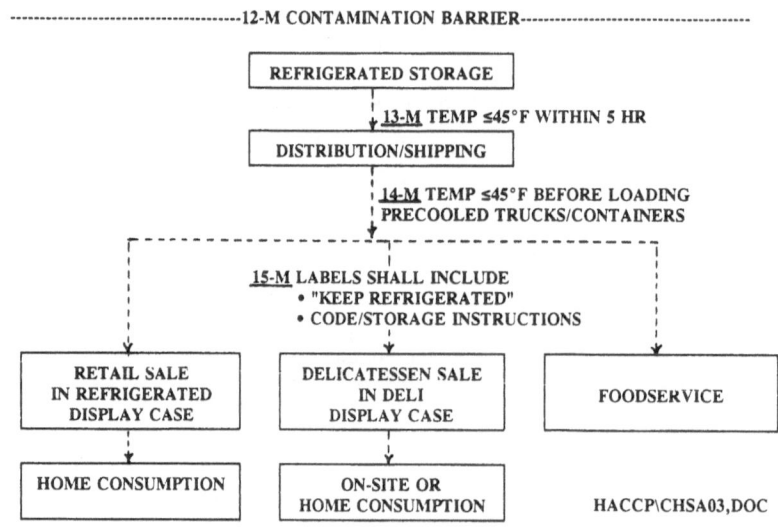

FIGURE 14.17.   Continued.

APPLICATION OF HACCP TO FOODSERVICE

An integral part of the development process for any food product is an evaluation of how new products could be used and abused. Some companies go so far as to invite consumers into their research centers for the express purpose of abusing their new ideas. The developers want to know if preparation instructions are clear; whether a raw product which requires cooking may be eaten without cooking; whether "easy open" packages are that; or whether consumers may hurt themselves or others when using a product. No matter what happens, liability ultimately lies with the company whose name is on the label. One of the more lucrative, yet difficult markets to manufacture for is foodservice. This industry serves enormous volumes of food each day, yet they rely on what might be the most poorly educated, most transient, and youngest work force in the whole food industry. These individuals handle every meal or item which is served. It is a situation ripe for problems, problems which are all too real. Todd (1989a, 1989b) estimates that only 5% of all foodborne illnesses may be traced to abusive industrial practices. The remaining 95% are associated with abusive practices in food service, restaurant, or home preparation of foods. Bryan (1990) reiterates this concern, concerns which are magnified when an outlet prepares foods "from scratch". He suggests that food service operations examine their process flows,

CCP Number	CCP DESCRIPTION	CRITICAL LIMIT(S) DESCRIPTION
1-MPC	**HAZARD CONTROLLED;** Microbiological, Physical and Chemical Point or Procedure: Incoming Inspection	1.1 Sanitary Condition
		1.2 Refrig. Material ≤45°F
		1.3 Frozen Material ≤32°F
		1.4 Vendor met all safety specifications before shipping
2-T	**HAZARD CONTROLLED:** Microbiological Point or Procedure Refrigerated Ingredient Storage	2.1 Material internal temperature not to exceed 45°F
		2.2 Calibrate temperature-measuring devices before shift
3-M	**HAZARD CONTROLLED:** Microbiological Point or Procedure: Sanaitation Requirements in - Preparation area - Staging area - Filling/Packaging area	3.1 Comply with USDA sanitation requirements
		3.2 Sanitation crew trained
		3.3 Each area must pass inspection
		before shift start-up
	HAZARD CONTROLLED: Point or Procedure: *Listeria*	3.4 Food contact surface Microbiological test
		3.5 Environmental area Microbiological tests
		(USDA Methodology for 3.4 & 3.5)
4-M	**HAZARD CONTROLLED:** Microbiological Point or Procedure: Controlled treatment to reduce microbiological contamination on raw celery and onions	Application of alternative approved treatments 4.1 Wash product with water containing - Chlorine, or - Iodine, or - Surfactants, or - No process additives
		4.2 Hot water or steam blanch followed by chilling
		4.3 Substitute processed celery or onions: - Blanched frozen - Blanched dehydrated - Blanched canned

FIGURE 14-18.   Critical limits for refrigerated chicken salad.

5-M	**HAZARD CONTROLLED:** Microbiological	5.1 Not to exceed 45°F
	Point or Procedure: Chilled storage temperature of prepared celery, onions and chicken	5.2 Refrigerator not to exceep 45°F
		5.3 Daily calibration of temperature measuring devices
6-MPC	**HAZARD CONTROLLED** Microbiological, Physical and Chemical	6.1 Physical barrier in-place
		6.2 Doors kept closed when not in use
	Point or Procedure: Physical barrier to prevent cross-contamination from raw material preparation area	6.3 Color-coded uniforms
		6.4 Supervision in-place
7-M	**HAZARD CONTROLLED:** Microbiological	7.1 Comply with USDA sanitation requirements
	Point or Procedure: Cross-contamination prevention from transfer equipment from raw material area	7.2 Prevent entry of soiled pallets cart wheels, totes, and other equipment
8-M	**HAZARD CONTROLLED:** Microbiological	8.1 Time limit not to exceed four hours for any materials in staging area
	Point or Procedure: Time limit for in-process food materials	
9-M	**HAZARD CONTROLLED:** Microbiological	9.1 Product pH must not exceed a pH of 5.5
	Point or Procedure: Maximum pH limit on finished salad before packaging	9.2 pH meter must be calibrated with approved standards before each shift
10-M	**HAZARD CONTROLLED;** Microbiological	10.1 Internal temperature not to exceed 45°F
	Point or Procedure: Chilled product storage temperature and time before packaging	10.2 Product must not be held more than one shift before filling/packaging
11-P	**HAZARD CONTROLLED:** Physical	11.1 Ferrous metal detection device for individual packages
	Point or Procedure: Metal detector for packages	11.2 Calibration or inspection not to exceed every four hours

FIGURE 14-18. Continued.

12-M	**HAZARD CONTROLLED:** Microbiological	12.1 Physical barrier in place
	Point or Procedure	12.2 Doors kept closed when not in use
	Physical barrier to prevent cross-contamination from warehouse area	12.3 Color coded uniforms
		12.4 Supervision in-place
13-M	**HAZARD CONTROLLED:** Microbiological Point or Procedure: Refrigerated storage of cased/ palleted finished product	13.1 Product internal temperature not to exceed 45°F in four hours
		13.2 Temperature measuring devices calibrated before shift
14-M	**HAZARD CONTROLLED:** Microbiological Point or Procedure: Truck and shipping containers for distribution of finished product	14.1 Shipping compartments must be pre-cooled to 45°F or less before loading product
15-M	**HAZARD CONTROLLED:** Microbiological Point or Procedure: Label Instructions	15.1 Each package or bulk case shall have label instructions
		15.2 Each laabel shall include: - Keep Refrigerated - Code - Storage Instructions

FIGURE 14-18.   Continued.

and implement HACCP programs. A potato salad prepared in-house is used as an example of how to set critical control points.

There are many foodservice outlets that prepare foods from raw materials, but what is becoming more common is the use of prepared items, which need only to be served or reheated before serving. Examples of such products are French fries, lettuce, chicken salad (three models used here), fried chicken, pastries, and specialty items. Mushrooms are commonly used on pizzas as toppings, but can be used as ingredients in salads or other products or presented as part of a salad bar. HACCP principles can be applied in foodservice operations, as Bryan (1990) implied. In fact, implementation of such programs in foodservice would probably go far towards reducing the number of outbreaks of foodborne illness. Let us take our models a bit further and determine CCPs for each at the foodservice level.

For each product, the first critical control point will be at the receiving dock or area. Those responsible for deliveries at stores, schools, restaurants, or markets must examine the condition of each product as it is unloaded. They should also take a look at the trucks, themselves. A filthy truck can be an indication that the products may have been abused and can also serve as a vehicle for infesting the kitchen itself with undesirable pests. Any suspect materials should be rejected at receipt.

When receiving canned mushrooms, look for damaged cases or cans which are swollen or leaking. Be sure the delivery truck is well maintained and that materials are supplied by a known and approved supplier.

For lettuce, chicken salad, and French fries, the first key is maintenance of refrigeration. It is recommended that all trucks have functional temperature indicators, which should be checked. Many products now are fitted with temperature indicators to monitor abuse. Look at these if they are used. Examine the product for evidence of temperature abuse, which is most easily seen with frozen products. Loss of temperature control frequently causes ice to accumulate on pallets or boxes. Finally, look at the condition of the delivery truck and be sure that the products are produced by approved suppliers.

Many foodservice operations (fast food and restaurant chains) have their own distribution system and truck fleets. These vehicles must be examined. Simply because a truck is part of your own company is not an excuse for failing to check incoming materials.

Mushrooms

Canned mushrooms are the safest of the four products. Chain restaurants or kitchens using this product need not be concerned about temperature control of the product. The only way to contaminate the canned product is to abuse it to the point that the cans leak and become recontaminated, which would be obvious. The major problem lies with opened cans in which only some of product has

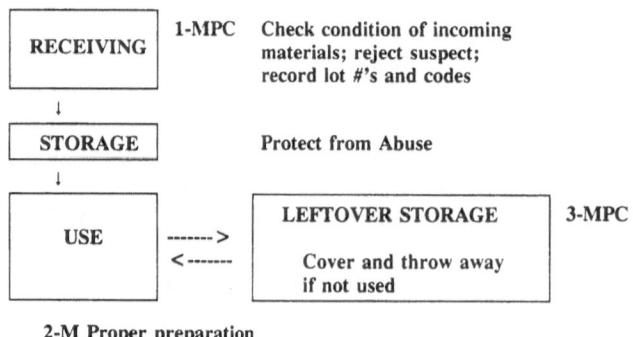

FIGURE 14-19.   Process flow for canned mushrooms in foodservice.

been used. These must be refrigerated and covered to protect them from recontamination. Operators should institute a policy where they retain records of lots or products codes used each day. If a problem does occur, be it illness or injury, the restaurant can set similarly coded materials aside for testing. A HACCP flow chart for this product is presented in Fig. 14-19.

Shredded lettuce

Lettuce is used in salad bars, as a condiment for burgers, in tacos, and in many other products in foodservice. It has not been, nor will it ever be subjected to a kill step. The product must be handled to prevent recontamination and refrigerated at all times. Finally, restaurant operators should not use sulfites to prevent browning, a procedure which was a standard practice in past years. If they do so, warnings must be posted. This can be dangerous since a small segment of the population is sensitive to the preservative. At this level, the key to product safety is sanitation and employee education. With the work force involved, the latter may be difficult, but it must be done. Figure 14-20 outlines a HACCP plan for the product.

For CCP 3-M, the lettuce must be kept cold, the temperatures monitored, and the area thoroughly cleaned and sanitized daily. Operators should not use sulfites, but if they do, usage levels must be controlled and warning signs posted. For CCP 4-M, workers handling product must use sanitary utensils for serving or applying and keep their hands and uniforms clean. These individuals should have received some basic education in food handling and sanitation before they are allowed to handle food. In an operation such as this, it is up to the manager to ensure that no workers who are sick or have open wounds are allowed to

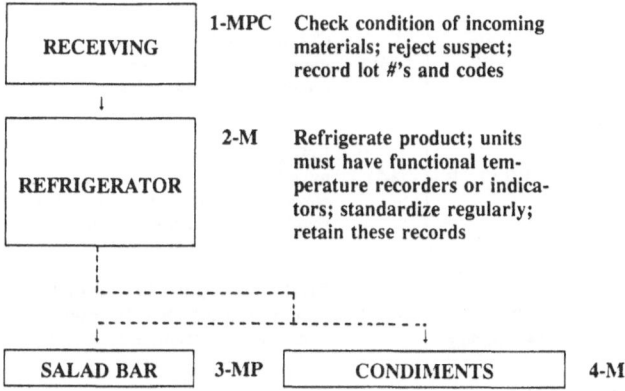

FIGURE 14-20.   Process flow for shredded lettuce in foodservice.

work. Finally, unused product should be thrown away at the end of the day, both for safety and quality reasons.

Chicken salad

Chicken salad has been implicated in a number of outbreaks of foodborne illnesses, as described earlier (Bryan 1988a). Obviously, there would be greater concerns were the product assembled in an institutional kitchen from the component materials, but this model uses the chicken salad produced commercially.

Foodservice uses both bulk and single serving chicken salads. The single serve item is easier and safer to handle in foodservice, but is more expensive. Bulk salads must be opened and served, or set out on a salad bar. Since the hazard potential is greater for the bulk salads, the process flow in Fig. 14-21 is based on that product. For simplicity sake, let us further assume that the salad is being served to individual customers.

If "leftover" salad is to be used, it must be returned to the refrigerator and protected. This additional handling step can create additional problems. As with the lettuce, managers should work to educate their staff about proper food handling and must prevent employees who are sick or who have open wounds (including boils or pimples) from handling food. Finally, food serving and holding areas must be cleaned and sanitized regularly.

French fries

The final model is French fries. There have probably not been too many illnesses or injuries caused by consumption of French fries over the years. It is a simple product and is generally eaten piping hot. Operators must still take steps to

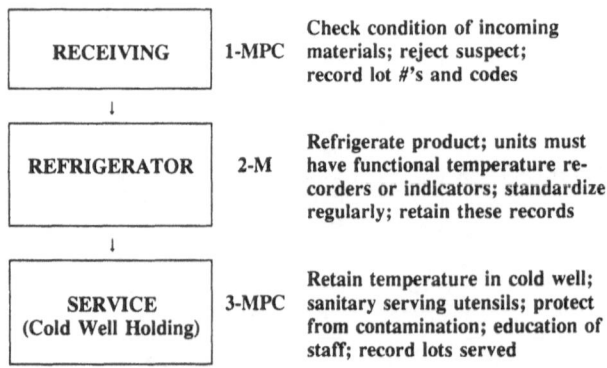

FIGURE 14-21.   Process flow for chicken salad in foodservice.

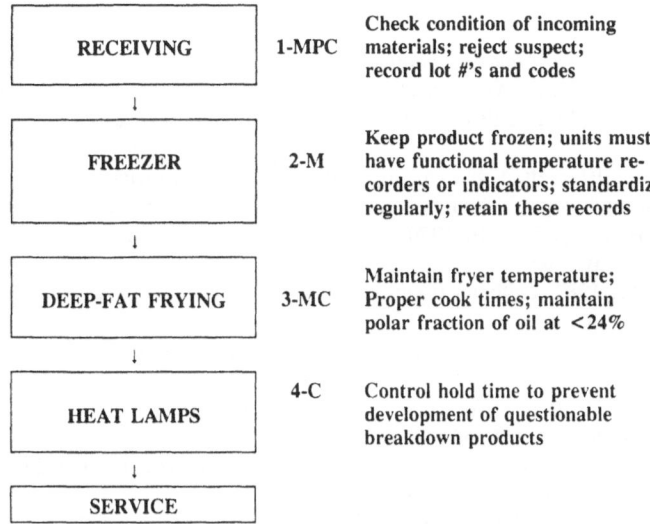

RECEIVING	1-MPC	Check condition of incoming materials; reject suspect; record lot #'s and codes
FREEZER	2-M	Keep product frozen; units must have functional temperature recorders or indicators; standardize regularly; retain these records
DEEP-FAT FRYING	3-MC	Maintain fryer temperature; Proper cook times; maintain polar fraction of oil at <24%
HEAT LAMPS	4-C	Control hold time to prevent development of questionable breakdown products
SERVICE		

FIGURE 14-22.   Process flow for French fries in foodservice.

ensure that the product served is safe. Figure 14-22 highlights what critical control points might be established for this product.

The frying process (or baking) will destroy any viable organisms on the French fries. Toxins (microbial or other) may not be destroyed, but control of the system upstream should prevent their even being present. The primary concern at the foodservice level with French fries is the frying oil. Food fried in badly abused oils may absorb the degraded fat, causing gastrointestinal distress. Complaints of this nature and studies on oil quality eventually led to the development of regulations governing restaurant frying oils in Europe (Firestone et al. 1991).

SUMMARY

The objective of this chapter was to provide individuals interested in HACCP with guidelines for use in their own operations. The focus was on the first three principles of HACCP, that is, risk assessment, determining critical control points, and setting limits. The remaining four principles will be left to the readers to develop for themselves.

ACKNOWLEDGMENTS

The author thanks friends and colleagues in the industry who assisted in developing models and helping to assure that they mirrored "real world" practices.

References

Anonymous. 1988. *The Almanac of the Canning, Freezing, Preserving Industries*. Edward R. Judge & Sons, Inc., Westminster, MD.

Anonymous. 1989. Morbidity and Mortality Weekly, Center for Disease Control, Atlanta, GA, June 22.

Berrang, M.E., Brackett, R.E., and Beuchat, L.R. 1989. Growth of *Listeria monocytogenes* on fresh vegetables under controlled atmosphere. J. Food Prot. 52: 702–705.

Blackstock, H. and Skiver, B. 1974. Potatoes nourish the Simplot processing empire. Food Engr. Nov.: 59.

Blumenthal, M. 1987. Optimun Frying: Theory and Practice. Monograph Series. Libra Laboratories, Piscataway, NJ.

Bryan, F.L. 1988a. Risks associated with foodborne pathogens and toxins. J. Food Prot. 51: 498–508.

Bryan, F.L. 1988b. Risks of practices, procedures and processes that lead to outbreaks of foodborne disease. J. Food Prot. 51: 663–673.

Bryan, F.L. 1990. Hazard analysis critical control point systems for retail food and restaurant operations. J. Food Prot. 53: 978–983.

Clark, W.L. and Serbia, G.W. 1991. Safety aspects of frying fats and oils. Food Technol. 45: 84–88, 94.

Corlett, D.A. and Stier, R.F. 1990. A practical application of HACCP. ESCAgenetics Corporation, San Carlos, CA.

Corlett, D.A. and Stier, R.F. 1991a. Risk assessment within the HACCP system. Food Control 2: 71–72.

Corlett, D.A. and Stier, R.F. 1991b. *A Practical Application of HACCP*. Short Course Manual from course presented for the Salad Manufacturers Association, Feb. 12–13, Atlanta GA.

Corlett, D.A. and Mitchell, 1991. HACCP model for refrigerated chicken salad. Submitted to USDA/FSIS, Feb. 28. (Unpublished)

Denny, C.B. 1982. Industry's response to problem solving in botulism prevention. Food Technol. 36: 116–118.

Firestone, D., Stier, R.F., and Blumenthal, M.M. 1991. Regulation of frying fats and oil. Food Technol. 45: 88–71, 94.

Hardt-English, P. York, G., Stier, R., and Cocotas, P. 1990. Staphyloccal food poisoning outbreaks caused by canned mushrooms from China. Food Techol. 44:74, 76–77.

Kallander, K.D., Witchkins, A.D., Lancette, G.A., Schmies, J.A., Garcia, G.A., Solomon, H.A., and Sofos, J.N. 1991. Fate of *Listeria monocytogenes* in shredded cabbage stored at 5 and 25°C under a modified atmosphere. J. Food Prot. 54: 302–304.

National Advisory Committee for Microbiological Criteria for Food (NACMCF) 1989. *HACCP Principles for Food Production*. USDA-FSIS Information Office Washington, DC.

National Food Processors Association (NFPA). 1989. Newsletter, Oct. 13.

National Oceanic and Atmospheric Administration/National Marine Fisheries Service (NOAA/NMFS). 1989a. HACCP Regulatory Model: Breaded Shrimp. NMFS Office of Trade and Industry Services, Pascagoula, MS.

National Oceanic and Atmospheric Administration/National Marine Fisheries Service (NOAA/NMFS). 1989b. HACCP Regulatory Model: Raw Shrimp. NMFS Office of Trade and Industry Services, Pascagoula, MS.

National Oceanic and Atmospheric Administration/National Marine Fisheries Service (NOAA/NMFS). 1990a. HACCP Regulatory Model: Blue Crab. NMFS Office of Trade and Industry Services, Pascagoula, MS.

National Oceanic and Atmospheric Administration/National Marine Fisheries Service (NOAA/NMFS). 1990b. HACCP Regulatory Model: Breaded Fish and Specialty Items. NMFS Office of Trade and Industry Services, Pascagoula, MS.

Reister, D. 1974. Whatever happened to canned mushrooms? FDA Consumer, July–August. Government Printing Office, Washington, DC.

Satchell, F.B., Stephenson, P., Andrews, W.H., Estela, L., and Allen, G. 1990. The survival of Shigella sonnei in shredded cabbage. J. Food Prot. 53: 558–562.

Sizmur, K. and Walker, C.W. 1988. Listeria in prepackaged salads. Lancet 1(8595): 1167.

Steinbrugge, E.S., Maxcy, R.B., and Liewen, M.B. 1988. Fate of Listeria monocytogenes on ready-to-serve lettuce. J. Food Prot. 51:596–599.

Todd, E.C.D. 1989a. Preliminary estimates of the cost of food-borne disease in Canada and costs to reduce Salmonella. J. Food Prot. 52:586–594.

Todd, E.C.D. 1989b. Preliminary costs of foodborne illness in the United States. J. Food Prot. 52:595–601.

U.S. Food & Drug Administration (FDA). 1989. Thermally processed low-acid foods packaged in hermetically sealed containers, Title 21, Part 113, Office of the Federal Register, Washington, DC.

A

HACCP

United States
Department of
Agriculture

Food Safety
and Inspection
Service

HACCP
Principles
for Food
Production

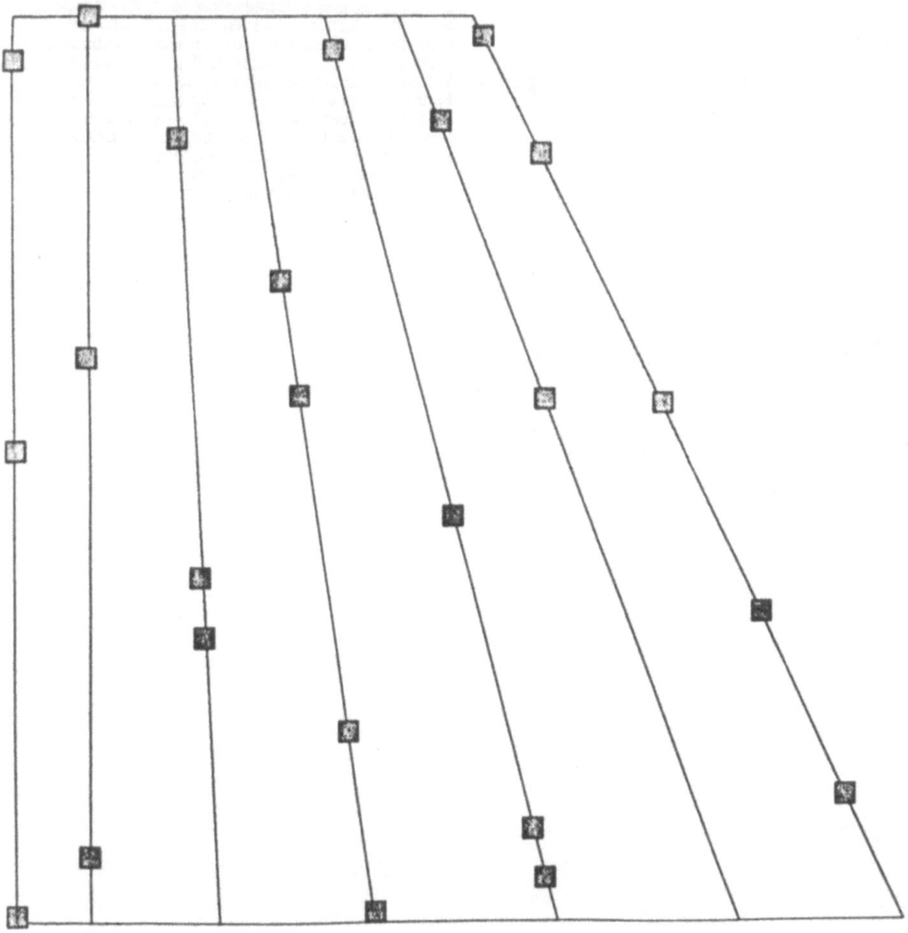

Hazard Analysis and Critical Control Point System
National Advisory Committee on Microbiological
Criteria for Foods

169

HAZARD ANALYSIS &
CRITICAL CONTROL POINT SYSTEM

NATIONAL ADVISORY COMMITTEE ON
MICROBIOLOGICAL CRITERIA FOR FOODS

ADOPTED
NOVEMBER 1989

TABLE OF CONTENTS

Executive Summary

In response to a request from the Chairman of the National Advisory Committee on Microbiological Criteria for Foods (Committee), Dr. Lester M. Crawford, an ad hoc working group chaired by Dr. Donald A. Corlett undertook the assignment of drafting a guide setting forth the principles of Hazard Analysis Critical Control Point (HACCP) Systems. The Committee has espoused HACCP as an effective and rational approach to the assurance of food safety. This document represents "HACCP" as used by this Committee. There has been no attempt made to draft a specific HACCP plan for any commodity. HACCP systems must be developed by individual producers and tailored to their individual processing and distribution conditions.

This document defines HACCP as a systematic approach to be used in food production as a means to assure food safety. Seven basic principles underlie the concept. These principles include an assessment of the inherent risks that may be present from harvest through ultimate consumption. Six hazard characteristics and a ranking schematic are used to identify those points throughout the food production and distribution system whereby control must be exercised in order to reduce or eliminate potential risks. A guide for HACCP plan development and critical control point (CCP) identification are noted. Further, the document points out the additional areas that are to be included in the HACCP plan—the need to establish critical limits that must be met at each CCP, appropriate monitoring procedures, corrective action procedures to be taken if a deviation is encountered, recordkeeping, and verification activities.

1.0 Preamble

The National Advisory Committee on Microbiological Criteria for Foods (Committee) endorses the Hazard Analysis and Critical Control Point (HACCP) System as an effective and rational approach to the assurance of food safety. In the application of HACCP, the use of microbiological testing is seldom an effective means of monitoring critical control points (CCP) because of the time required to obtain results. In most instances, monitoring of CCP can best be accomplished through the use of physical and chemical tests, and through visual observations. Microbiological criteria do, however, play a role in verifying that the overall HACCP system is working.

The Committee believes that the HACCP principles should be standardized to create uniformity in its work, and in training and applying the HACCP system by industry and regulatory authorities. In accordance with the National Academy of Sciences recommendation, the HACCP system must be developed by each food establishment and tailored to its individual product, processing and distribution conditions.

2.0 Definitions

2.1   Continuous Monitoring: Uninterrupted record ing of data such as a recording of temperature on a strip chart.

2.2   Control Point: Any point in a specific food system where loss of control does not lead to an unacceptable health risk.

2.3   Critical Control Point: Any point or procedure in a specific food system where loss of control may result in an unacceptable health risk.

2.4   Critical Defect: A defect that may result in hazardous or unsafe conditions for individuals using and depending upon the product.

2.5   Critical Limit: One or more prescribed toler ances that must be met to insure that a critical control point effectively controls a microbio logical health hazard.

2.6   Deviation: Failure to meet a required critical limit for a critical control point.

2.7   HACCP Plan: The written document which delineates the formal procedures to be followed in accordance with these general principles.

2.8   HACCP System: The result of the implementa tion of the HACCP principles.

2.9   Hazard: Any biological, chemical, or physical property that may cause an unacceptable consumer health risk.

2.10   Monitoring: A planned sequence of observa tions or measurements of critical limits designed to produce an accurate record and intended to insure that the critical limit maintains product safety.

2.11   Risk: An estimate of the likely occurrence of a hazard or danger.

2.12   Risk Category: One of six categories prioritiz ing risk based on food hazards.

2.13   Sensitive Ingredient: Any ingredient histori cally associated with a known microbiological hazard.

2.14   Significant Risk: Posing moderate likelihood of causing an unacceptable health risk.

2.15   Spot Check: Supplemental tests performed on a random basis.

2.16   Verification: Methods, procedures and tests used to determine if the HACCP system is in compliance with the HACCP plan.

3.0 Purpose and Principles

HACCP is a systemic approach to food safety, consisting of the seven following principles:

3.1   Assess hazards and risks associated with growing, harvesting, raw materials and ingredi ents, processing, manufacturing, distribution, marketing, preparation and consumption of the food.

3.2   Determine CCP required to control the identi fied hazards.

3.3   Establish the critical limits that must be met at each identified CCP.

3.4   Establish procedures to monitor CCP.

3.5   Establish corrective action to be taken when there is a deviation identified by monitoring a CCP.

3.6   Establish effective record-keeping systems that document the HACCP plan.

3.7   Establish procedures for verification that the HACCP system is working correctly.

4.0 Explanation of Principles

4.1   Principle No. 1: Assess hazards associated with growing, harvesting, raw materials and ingredients, processing manufacturing, distribu tion, marketing, preparation and consumption of the food.

4.1.1   Description: Provides for a systematic evalu ation of a specific food and its ingredients or components to determine the risk from hazard ous microorganisms or their toxins. Hazard analysis is most useful for guiding the safe design of a food product and defining the CCP that eliminate or control hazardous microorgan isms or their toxins at any point during the entire production sequence. The hazard assessment is a two-part process consisting of

3

172 HACCP: Principles and Application

ranking a food according to six hazard characteristics, followed by the assignment of risk category which is based upon the ranking.

Ranking according to hazard characteristics is based on assessing a food in terms of (a) whether the product contains microbiologically sensitive ingredients, (b) whether the process does not contain a controlled processing step that effectively destroys harmful microorganisms, (c) whether there is significant risk of post processing contamination with harmful microorganisms or their toxins, and (d) whether there is substantial potential for abusive handling in distribution or in consumer handling or preparation that could render the product harmful when consumed or (e) whether there is no terminal heat process after packaging or when cooked in the home. Ranking according to these six characteristics results in the assignment of risk categories based on how many of the characteristics are present.

The risk categories are utilized for recognizing the hazard risk for ingredients and how they must be treated or processed to reduce the risk for the entire food production and distribution sequence.

The hazard assessment procedure is ideally conducted after developing a working description of the product, establishing the types of raw materials and ingredients required for preparation of the product, and preparing a diagram for the food production sequence. The two-part assessment of hazard analysis and assignment of risk categories is conducted according to the following procedure:

4.1.2   Hazard analysis and assignment of risk categories:

4.1.2.1 Rank the food according to hazard characteristics A through F, using a plus (+) to indicate a potential hazard. The number of pluses will determine the risk category. A model diagram outlining this concept is given under section 4.1.3. As indicated, if the product falls under hazard class A, it should automatically be considered Risk Category VI.

Hazard A:

A special class that applies to non-sterile products designated and intended for consumption by at risk populations, e.g., infants, the aged, the infirm, or immunocompromised individuals.

Hazard B:

The product contains "sensitive ingredients" in terms of microbiological hazards.

Hazard C:

The process does not contain a controlled processing step that effectively destroys harmful microorganisms.

Hazard D:

The product is subject to recontamination after processing before packaging.

Hazard E:

There is substantial potential for abusive handling in distribution or in consumer handling that could render the product harmful when consumed.

Hazard F:

There is no terminal heat process after packaging or when cooked in the home.

Note: Hazards can also be stated for chemical or physical hazards, particularly if a food is subject to them.

4.1.2.2 Assignment of risk category (based on ranking by hazard characteristics):

Category VI.

A special category that applies to non-sterile products designated and intended for consumption by at risk populations, e.g., infants, the aged, the infirm, or immunocompromised individuals. All six hazard characteristics must be considered.

Category V.

Food products subject to all five general hazard characteristics.
Hazard class B, C, D, E, F

Category IV.

Food products subject to four general hazard characteristics.

Category III.

Food products subject to three of the general hazard characteristics.

Category II.

Food products subject to two of the general hazard characteristics.

Category I.

Food products subject to one of the general hazard characteristics.

Category O.

Hazard Class—No hazard.

Note: Ingredients are treated in the same manner in respect to how they are received at the plant, BEFORE processing. This permits determination of how to reduce risk in the food system.

4.1.3   It is recommended that a chart be utilized that provides assessment of a food by hazard characteristic and risk category. A format for this chart is given as follows:

Food Ingredient or Product	Hazard Characteristics (A,B,C,D,E,F)	Risk Category
T	A+ (Special Category)*	VI
U	Five +'s (B through F)	V
V	Four +'s (B through F)	IV
W	Three +'s (B through F)	III
X	Two +'s (B through F)	II
Y	One + (B through F)	I
Z	No +'s	O

Hazard characteristic A automatically is risk category VI, but any combination of B through F may also be present.

4.2   Principle No. 2: Determine CCP required to control the identified hazards.

4.2.1   Description: A CCP is defined as any point or procedure in a specific food system where loss of control may result in an unacceptable health risk. CCP must be established where control can be exercised. All hazards identified by the hazard analysis must be controlled at some point(s) in the food production sequence, from harvesting and growing raw materials to the ultimate consumption of the food.

CCP are located at any point in a food sequence where hazardous microorganisms need to be destroyed or controlled. For example, a specified heat process, at a given time and temperature to destroy a specified microbiological pathogen, is a CCP. Likewise, refrigeration required to prevent hazardous organisms from growing, or the adjustment of a food to a pH necessary to prevent toxin formation is a CCP.

Types of CCP may include, but are not limited to: cooking, chilling, sanitizing, formulation control, prevention of cross contamination, employee hygiene and environmental hygiene.

CCP must be carefully developed and documented. In addition, they must be used only for purposes of product safety. They should not be confused with control points that do no control safety. For comparison, a control point is defined as any point in a specific food system where loss of control does not lead to an unacceptable health risk.

4.3   Principle No. 3: Establish the critical limits which must be met at each identified CCP.

4.3.1   Description: A critical limit is defined as one or more prescribed tolerances that must be met to insure that a CCP effectively controls a microbiological health hazard. There may be more than one critical limit for a CCP. If any one of those critical limits is out of control, the CCP will be out of control and a potential hazard can exist. The criteria most frequently utilized for critical limits are temperature, time, humidity, moisture level (Aw), pH, titratable acidity, preservatives, salt concentration, available chlorine, viscosity and in some cases, sensorial information such as texture, aroma and visual appearance. Many different types of limit information may be needed for safe control of a CCP.

For example, the cooking of meat patties should be designed to eliminate the most heat-resistant vegetative pathogen which could reasonably be expected to be in the product. The critical limits must be specified for temperature, time and meat patty thickness. Technical development of these critical limits requires accurate information on the probable maximum numbers of these microorganisms in the meat, use of additional ingredients and the potential for re-contamination.

The relationship between the CCP and its critical limits for the meat patty example is shown as follows:

Critical Control Point	Critical Limits
Meat patty cooked to destroy the most heat resistant pathogen, based on lethality tests. The minimum lethal cook will usually be designated "to reach an internal patty temperature of x for time y."	Minimum operating temperature of cooker to achieve microbiological lethality at center of coldest patty. Time to achieve lethality (belt speed expressed at rpm). Patty thickness. Other possible critical limits: —oven humidity —patty composition —cooker sanitation —etc.

This example illustrates that the type and number of critical limits will vary depending on the type of cooking system and equipment used for meat patties.

4.4   Principle No. 4 Establish procedures to monitor CCP.

4.4.1   Description: Monitoring is the scheduled testing or observation of a CCP and its limits. Monitoring results must be documented. From the monitoring standpoint, failure to control a CCP is a critical defect.

A critical defect is defined as a defect that may result in hazardous or unsafe conditions for individuals using and depending upon the product. Because of the potentially serious

5

consequences of a critical defect, monitoring procedures must be extremely effective.

Ideally, monitoring should be at the 100% level. Continuous monitoring is possible with many types of physical and chemical methods. For example, the temperature and time for the scheduled thermal process of low-acid canned foods is recorded continuously on temperature recording charts. If the temperature falls below the scheduled temperature or the time is insufficient, as recorded on the chart, the retort load is restrained as a process deviation. Likewise, pH measurement may be done continually in fluids or by testing of a batch before processing. There are many ways to monitor CCP limits on a continuous or batch basis and record the data on charts. The high reliability of continuous monitoring is always preferred when feasible. It requires careful calibration of equipment.

When it is not possible to monitor a critical limit on a full-time basis, it is necessary to establish that the monitoring interval will be reliable enough to indicate that the hazard is under control. Statistically designed data collection systems or sampling systems lend themselves to this purpose. However, statistical procedures are most useful for measuring and reducing the variation in food formulations, manufacturing equipment and measuring devices. Thus, they increase the reliability of the system.

When using statistical process control, it is important to recognize that there is no tolerance for exceeding a critical limit. For example, when a pH of 4.6 or less is required for product safety, no single product unit may have a pH above 4.6. To compensate for variation, the maximum of the product may be targeted at a pH below 4.6. Statistical process control can be applied to understand variation in the system, and assure that no unit exceeds a pH of 4.6. Statistical audits can be based on this concept.

Most monitoring procedures for CCP will need to be done rapidly because they relate to on-line processes and there will not be time for lengthy analytical testing. Microbiological testing is seldom effective for monitoring CCP due to their time-consuming nature. Therefore, physical and chemical measurements are preferred because they may be done rapidly and can indicate microbiological control of the process.

Physical and chemical measurements that may be utilized for monitoring include:

Temperature;
Time;
pH;
Sanitations at CCP;
Specific preventive measures for cross contamination;
Specific food handling procedures;
Moisture level; and
Other.

6

Spot checks are useful for supplementing the monitoring of certain CCP and their respective limits. They may be used to check incoming pre-certified ingredients, assess equipment and environmental sanitation, airborne contamination, cleaning and sanitizing of gloves and any place where follow-up is needed. Spot checks may consist of physical and chemical tests and, where needed, microbiological tests.

With certain foods, microbiologically sensitive ingredients, or imports, there may be no alternative to microbiological testing. However, a sampling frequency that is adequate for reliable detection of low levels of pathogens is seldom possible because of the large number of samples needed. For this reason, microbiological testing has limitations in a HACCP system, but is valuable as a means of establishing and randomly verifying the effectiveness of control at CCP, (challenge tests, spot checking or for troubleshooting.)

All records and documents associated with CCP monitoring must be signed by the person doing the monitoring and signed by a responsible official of the company.

4.5   **Principle No. 5:** Establish corrective action to be taken when there is a deviation identified by monitoring of a CCP.

4.5.1.   **Description:** Actions taken must eliminate the actual or potential hazard which was created by deviation from the HACCP plan, and assure safe disposition of the product involved. Because of the variations in CCP for different food and the diversity of possible deviations, specific corrective actions must be developed for each CCP in the HACCP plan. The actions must demonstrate that the CCP has been brought under control. Deviation procedures must be documented in the HACCP plan and agreed to by the appropriate regulatory agency prior to approval of the plan.

Should a deviation occur, the plant will place the product on hold pending completion of appropriate corrective actions and analyses. In instances where it may be difficult to determine the safety of the product, then the testing and final disposition must be agreed to by the government. In instances not associated with safety, government consultation is not required.

Identification of deviant lots and corrective actions taken to assure safety of these lots must be noted in the HACCP record and remain on file for a reasonable period after the expiration date or expected shelf life of the product.

4.6   **Principle No. 6:** Establish effective record-keeping systems that document the HACCP plan.

4.6.1   **Description:** The HACCP plan must be on file at the food establishment. Additionally, it is to include documentation relating to CCP and any action on critical deviations and disposition of product. These materials are to be made available to government inspectors upon request. The HACCP plan clearly designates records that will be available for government inspection. Certain records that deal with the functioning of the HACCP system and proprietary information are not necessarily available to regulatory agencies.

Generally, the types of records utilized in the total HACCP system will include the following:

(Note: Only those records pertaining to CCP must be made available to regulatory agencies.)

4.6.1.1 Ingredients

■ Supplier certification documenting compliance with processor's specifications.

■ Processor audit records verifying supplier compliance.

■ Storage temperature record for tem perature sensitive ingredients.

■ Storage time records of limited shelf life ingredients.

4.6.1.2 Records relating to product safety

■ Sufficient data and records to establish the efficacy of barriers in maintaining product safety.

■ Sufficient data and records establishing the safe shelf life of the product.

■ Documentation of the adequacy of the processing procedures from a knowledgeable process authority.

4.6.1.3 Processing

■ Records from all monitored CCP.

■ System records verifying the continued adequacy of the processes.

4.6.1.4 Packaging

■ Records indicating compliance with specifications of packaging materials.

■ Records indicating compliance with sealing specifications.

4.6.1.5 Storage and Distribution

■ Temperature records.

■ Records showing no product shipped after shelf life date on temperature sensitive products.

4.6.1.6 Deviation File

4.6.1.7 Modification to the HACCP plan file indicating approved revisions and changes in ingredients, formulations, processing, packaging and distribution control, as needed.

4.7 **Principle No. 7:** Establish procedures for verification that the HACCP system is working correctly.

4.7.1 **Description:** Verification consists of methods, procedures and tests used to determine that the HACCP system is in compliance with the HACCP plan. Both the producer and the regulatory agency have a role in verifying HACCP plan compliance. Verification confirms that all hazards were identified in the HACCP plan when it was developed. Verification measures may include physical, chemical and sensory methods and testing for conformance with microbiological criteria when established.

4.7.1.1 Examples of verification activities include but are not limited to:

■ Establishment of appropriate verification inspection schedules.

■ Review of the HACCP plan.

■ Review the CCP records.

■ Review deviations and dispositions.

■ Visual inspections of operations to observe if CCP are under control.

■ Random sample collection and analysis.

■ Written record of verification inspections which certifies compliance with the HACCP plan or deviations from the plan and the corrective actions taken.

4.7.1.2 Verification inspections should be conducted when:

■ Routinely, or on an unannounced basis to assure selected CCP are under control.

■ It is determined that intensive coverage of a specific commodity is needed because of new information on food safety issues requiring assurance that the HACCP plan remains effective.

■ Foods produced have been implicated as a vehicle of foodborne disease.

■ Requested on a consultative basis or established criteria have not been met.

7

4.7.1.3 Elements which must be included in verification inspection reports:

■ Existence of an approved HACCP plan and designation of person(s) responsible for administering and updating the HACCP plan.

■ All records and documents associated with CCP monitoring must be signed by the person monitoring and approved by a responsible official of the firm.

--Direct monitoring data of the CCP while in operation.

--Certification that monitoring equipment is properly calibrated and in working order.

--Deviation procedures.

■ Any sample analysis for attributes confirming that CCP are under control to include physical, chemical, microbiological or organoleptic methods.

5.0   Guide for HACCP plan Implementation for a specific food:

5.1   Describe the food and its intended use.

5.2   Develop a flow diagram for the production of the food.

5.3   Perform a hazard assessment (Principle 1).

a. Ingredients prior to any processing step.

b. End product.

5.4   Select CCP (Principle 2).

a. Enter on flow diagram in numerical order.

b. List CCP number and description.

5.5   Establish critical limits (Principle 3).

5.6   Establish monitoring requirements (Principle 4).

5.7   Establish corrective action to be taken when there is a deviation identified by monitoring of a CCP (Principle 5).

5.8   Establish effective recordkeeping systems that document the HACCP plan (Principle 6).

5.9   Establish procedures for industrial and governmental verification that the HACCP system is working properly. Verification measure may include physical, chemical and sensory methods, and when needed, establishment of microbiological criteria (Principle 7).

B

Codex Committee on Food Hygiene *DRAFT* HACCP Principles

At their June 14, 1991 meeting, the Codex Committee on Food Hygiene HACCP Drafting Group developed a *draft* report on HACCP. Subsequently, in October 1991, the Codex Committee on Food Hygiene made minor revisions and requested that the report be circulated to member countries for comments. It is anticipated that the HACCP document will be finalized in 1993. The Drafting Group recommended seven principles (similar to those of the National Advisory Committee on Microbiological Criteria for Foods), application tasks for the principles, and a logic sequence for application of HACCP. It is important to note that at the time of preparing this book the report given in this appendix was a *DRAFT* document and not finalized.

DRAFT PRINCIPLES AND APPLICATION OF THE HAZARD ANALYSIS CRITICAL CONTROL POINT (HACCP) SYSTEM

PREAMBLE

The Hazard Analysis Critical Control Point (HACCP) system offers considerable benefits for food safety. Therefore, the Codex Committee on Food Hygiene recommended that its use should be encouraged. The purpose of this document is to state the principles to be used in applying HACCP to foods and outline its application, with special reference to Codex codes of practice and standards.

HACCP is primarily applied by the food industry, but is equally applicable throughout the food chain from the primary producer to final consumer. As well as enhanced food safety, benefits include better use of resources and more timely response to problems. In addition, the application of HACCP systems can aid inspection by regulatory authorities and promote international trade by increasing confidence in food safety.

HACCP is a system which identifies specific hazards and preventative measures for their control to ensure the safety of food. HACCP is a tool to assess hazards and establish control systems that focus on preventative measures rather than relying mainly on end-product testing. Any HACCP system is capable of accommodating change, such as advances in equipment design, processing procedures or technological developments.

HACCP's successful application requires the full commitment and involvement of management and the workforce. It also requires a team approach; this team should include appropriate experts such as agronomists, veterinarians, production personnel, micro-biologists, medical experts, public health specialists, chemists and engineers according to the particular study. The application of HACCP is compatible with the implementation of quality management systems, such as the ISO 9000 series, and is the system of choice in the management of food safety within such systems.

While the application of HACCP to food safety was considered here, the system can equally be applied to other aspects of food quality.

PRINCIPLES

HACCP is a system which identifies specific hazard(s) (i.e., any biological, chemical or physical property that adversely affects the safety of the food) and preventative measures for their control. The system consists of the following seven principles:

Principle 1

Identify the potential hazard(s) associated with food production at all stages, from growth, processing, manufacture and distribution, until the point of consumption. Assess the likelihood of occurrence of the hazard(s) and identify the preventative measures for their control.

Principle 2

Determine the points/procedures/operational steps that can be controlled to eliminate the hazard(s) or minimize its likelihood of occurrence—(Critical Control Point (CCP)). A "step" means any stage in food production and/or manufacture including raw materials, their receipt and/or production, harvesting, transport, formulation, processing, storage, etc.

Principle 3

Establish target level(s) and tolerances which must be met to ensure the CCP is under control.

Principle 4

Establish a monitoring system to ensure control of the CCP by scheduled testing or observations.

Principle 5

Establish the corrective action to be taken when monitoring indicates that a particular CCP is not under control.

Principle 6

Establish procedures for verification which includes supplementary tests and procedures to confirm that HACCP is working effectively.

Principle 7

Establish documentation concerning all procedures and records appropriate to these principles and their application.

APPLICATION OF THE PRINCIPLES
OF HACCP

During the hazard analysis and subsequent operations in designing and applying HACCP systems, consideration must be given to the impact of raw materials, ingredients, food manufacturing practices, role of manufacturing processes to control hazards, likely end-use of the product, consumer populations at risk and epidemiological evidence relative to food safety.

The intent of the HACCP system is to focus control at CCPs. Redesign of the operation should be considered if a hazard is identified but no CCPs are found. HACCP should be applied to each specific process separately. CCPs identified in an example of a process might not be the only ones identified for a specific application or might be of a different nature. HACCP systems should be developed for specific processes.

The HACCP application should be reviewed and necessary changes made when any modification is made in the product, process, or any step.

It is important when applying HACCP to be flexible given the context of the application.

Application

The application of HACCP principles requires the following tasks as identified in the Logic Sequence for Application of HACCP (Figure B-1).

1. *Assemble HACCP team*
   Assemble a multidisciplinary team that has specific knowledge and expertise appropriate to the product. Where such expertise is not available on site, expert advice should be obtained from other sources.
2. *Describe product*
   A full description of the product should be drawn up including information on composition and method of distribution.

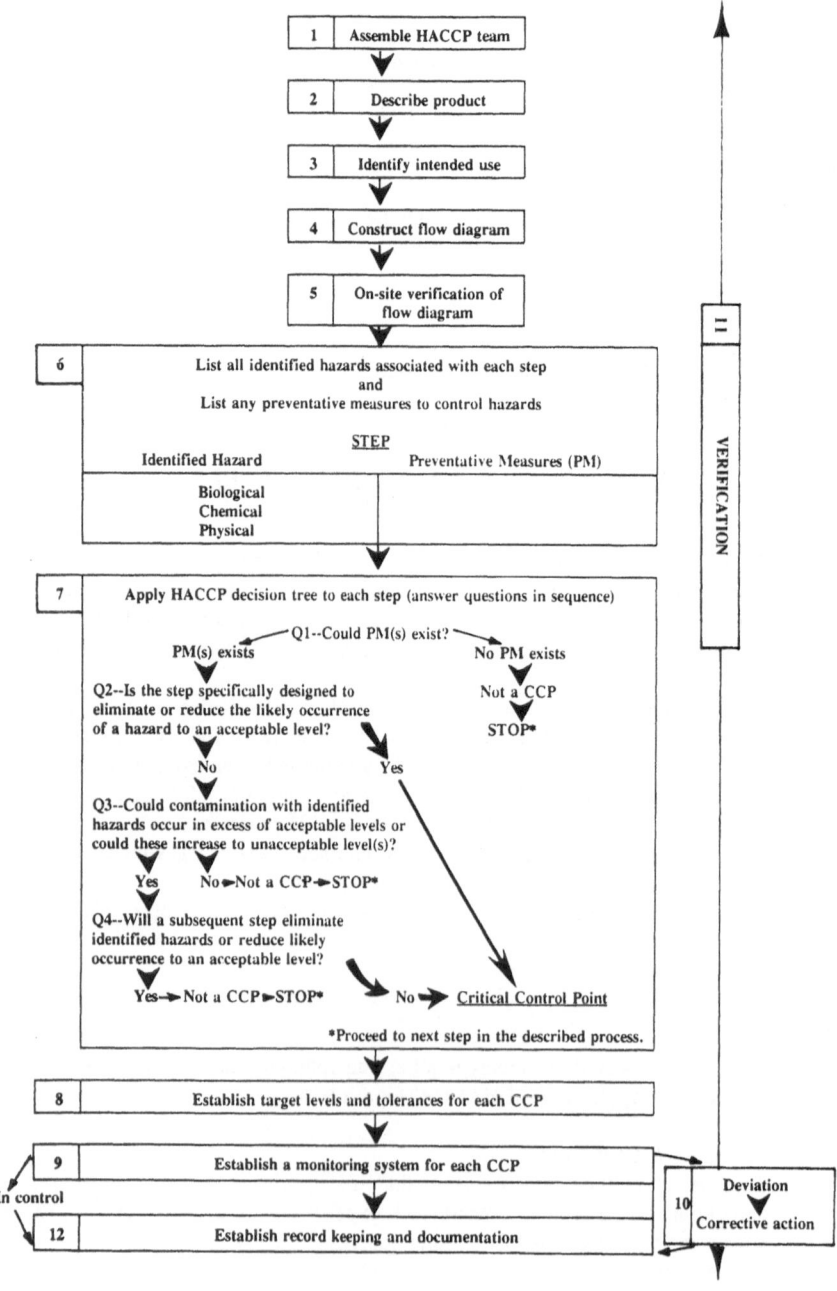

FIGURE B-1.   Logic sequence for application of HACCP.

3. *Identify intended use*
   The intended use should be based on the expected uses of the product by the end user or consumer. In specific cases, vulnerable groups of the population, e.g., institutional feeding, may have to be considered.
4. *Construct flow diagram*
   The flow diagram should be constructed by the HACCP team. Each step within the specified area of operation should be monitored and audited to produce the flow diagram. The flow diagram should be constructed for the particular part of the operation under consideration. When applying HACCP to a given operation, consideration should be given to steps preceeding and following the specified operation.
5. *On-site verification of flow diagram*
   The HACCP team should confirm the processing operation against the flow diagram during all stages and hours of operation and amend the flow diagram where appropriate.
6. *List all hazards associated with each step and list any preventative measures to control hazards (Principle 1)*
   The HACCP team should list all the biological, chemical or physical hazards that may be reasonably expected to occur at each step and describe the preventative measures that can be used to control these hazards.
   For inclusion in the list, hazards must be of a nature such that their elimination or reduction to acceptable levels is essential to the production of a safe food.
   Preventative measures are those actions and activities that are required to eliminate hazards or reduce their impact or occurrence to acceptable levels. More than one preventative measure may be required to control a specific hazard(s) and more than one hazard may be controlled by a specified preventative measure.
7. *Apply HACCP Decision Tree to Each Step (Principle 2)*
   The identification of a CCP in the HACCP system requires the application of a decision tree (Figure B-1). All hazards that may be reasonably expected to occur, or be introduced at each step, should be considered. Training in the application of decision trees may be required.
   If an identified hazard has no preventative measure at the step then no CCP exists at the step.
   Application of the decision tree determines whether the step is a CCP for the identified hazard. Application of the decision tree should be flexible, given whether the operation is for production, slaughter, processing, storage, distribution or other.
8. *Establish target levels and tolerances for each CCP (Principle 3)*
   Target levels and tolerances must be specified for each preventative measure. In some cases more than one target level and tolerance will be elaborated at a particular step. Criteria often used include measurements of temperature, time, moisture level, pH, Aw, and available chlorine, and organoleptic parameters such as visual appearance and texture.
9. *Establish a Monitoring System for Each CCP (Principle 4)*
   Monitoring is the scheduled measurement or observation of a CCP relative to its target levels and tolerances. The monitoring procedures must be able to detect

loss of control at the CCP. Further, monitoring should ideally provide this information in time for corrective action to be taken to regain control of the process before there is a need to reject the product. Data derived from monitoring must be evaluated by a designated person with knowledge and authority to carry out corrective actions when indicated. If monitoring is not continuous, then the amount or frequency of monitoring must be sufficient to guarantee the CCP is in control. Most monitoring procedures for CCPs will need to be done rapidly because they relate to on-line processes and there will not be time for lengthy analytical testing. Physical and chemical measurements are often preferred to microbiological testing because they may be done rapidly and can often indicate the microbiological control of the product. All records and documents associated with monitoring CCPs must be signed by the person(s) doing the monitoring and by a responsible reviewing official(s) of the company.

10. *Establish Corrective Actions (Principle 5)*
Specific corrective actions must be developed for each CCP in the HACCP system in order to deal with deviations when they occur.

The actions must ensure that the CCP has been brought under control. Actions taken must also include proper disposition of the affected product. Deviation and product disposition procedures must be documented in the HACCP record keeping.

Corrective action should also occur when monitoring results indicate a trend towards loss of control at a CCP. Action should be taken to bring the process back into control before the deviation leads to a safety hazard.

11. *Verification (Principle 6)*
Establish procedures for verification that the HACCP system is working correctly. Monitoring and auditing methods, procedures and tests, including random sampling and analysis, can be used to determine if the HACCP system is working correctly. Examples of verification activities include:
    Review of the HACCP system and its records.
    Review of deviations and product dispositions.
    Operations to observe if CCPs are under control.
    Validation of established target levels and tolerances.
    The frequency of verification should be sufficient to validate the HACCP system.

12. *Establish Record Keeping and Documentation (Principle 7)*
Efficient and accurate record keeping is essential to the application of a HACCP system. Documentation of HACCP procedures at all steps should be included and assembled in a manual.
    Examples of records are:
        Ingredients
        Records relating to product safety
        Processing
        Packaging
        Storage and distribution
        Deviation file
        Modifications to the HACCP system
    A convenient model checklist is attached as Figure B-2

HACCP CHECKLIST

1. | Describe product |

2. | Diagram process flow |

3. List

Step	Hazards	Preventative Measures	CCP	Target Level and Tolerance	Monitoring Procedures	Corrective Actions

FIGURE B-2.   HACCP checklist.

TRAINING

Training of personnel in industry, government and academia in HACCP principles and applications, and increasing awareness of consumers are essential elements for the effective implementation of HACCP. The International Commission on Microbiological Specifications for Foods (ICMSF) Monograph, "HACCP in Microbiological Safety and Quality," which describes the type of training required for various target groups, is recommended as a general approach to training (Blackwell Scientific Publications, Oxford Mead, UK, 1988, reprinted 1989). The section on training (Chapter 8) in the above monograph is equally applicable as an approach to training in respect to hazards other than those of microbiological nature.

Cooperation between industry, trade groups, consumer organizations and responsible authorities is of vital importance. Opportunities should be provided for the joint training of industry and control authorities to encourage and maintain a continuous dialogue and create a climate of understanding in the practical application of HACCP.

C

National Advisory Committee on Microbiological Criteria for Foods, Hazard Analysis and Critical Control Point System Adopted March 20, 1992

On March 20, 1992 the National Advisory Committee on Microbiological Criteria for Foods (NACMCF) adopted a revised document on "Hazard Analysis and Critical Control Point System." This was a revision of the NACMCF's 1989 document titled "HACCP Principles for Food Production."

HAZARD ANALYSIS AND CRITICAL CONTROL POINT SYSTEM

NATIONAL ADVISORY COMMITTEE ON MICROBIOLOGICAL CRITERIA FOR FOODS

March 20 1992

EXECUTIVE SUMMARY

The National Advisory Committee on Microbiological Criteria for Foods reconvened a Hazard Analysis and Critical Control Point (HACCP) Working Group in July 1991. The primary purpose of the working group was to review the Committee's November 1989 HACCP document comparing it with a draft report prepared by a HACCP Working Group of the Codex Committee on Food Hygiene. Based upon its review, the Committee has determined to expand upon its initial report by emphasizing the concept of prevention, incorporating a decision tree intended to facilitate the identification of Critical Control Points (CCPs), and providing a more detailed explanation of the application of HACCP principles.

The Committee again endorses HACCP as an effective and rational means of assuring food safety from harvest to consumption. Preventing problems from occurring is the paramount goal underlying any HACCP system. Seven basic principles are employed in the development of HACCP plans that meet the stated goal. These principles include hazard assessment, CCP identification, establishing critical limits, monitoring procedures, corrective actions, documentation, and verification procedures. Under such systems, if a deviation occurs indicating that control has been lost, the deviation is detected and appropriate steps are taken to re-establish control in a timely manner to assure that potentially hazardous products do not reach the consumer.

In the application of HACCP, the use of microbiological testing is seldom an effective means of monitoring critical control points (CCPs) because of the time required to obtain results. In most instances, monitoring of CCPs can best be accomplished through the use of physical and chemical tests, and through visual observations. Microbiological criteria do, however, play a role in verifying that the overall HACCP system is working.

The Committee believes that the HACCP principles should be standardized to create uniformity in its work, and in training and applying the HACCP system by industry and regulatory authorities. In accordance with the National Academy of Sciences recommendation, the HACCP system must be developed by each food establishment and tailored to its individual products, processing and distribution conditions.

In keeping with its charge of providing recommendations to its sponsoring agencies regarding microbiological food safety issues, this document focuses on microbiological safety. The Committee recognizes that in order to assure food safety, properly designed HACCP systems must also consider chemical and physical hazards in addition to other microbiological hazards.

In order for a successful HACCP program to be implemented, management must be committed to a HACCP approach. A commitment by management will indicate an awareness of the benefits and costs of HACCP and include education and training of employees. Benefits, in addition to food safety, are better use of resources and timely response to problems.

The Committee designed this document to guide the food industry in the implementation of HACCP systems. The Committee recommends that future documents address the role of regulatory agencies in the HACCP system.

1.0 DEFINITIONS

1.1 *CCP Decision Tree*: A sequence of questions to determine whether a control point is a CCP.

1.2 *Continuous Monitoring*: Uninterrupted collection and recording of data such as temperature on a strip chart.

1.3 *Control*: (a) To manage the conditions of an operation to maintain compliance with established criteria. (b) The state wherein correct procedures are being followed and criteria are being met.

1.4 *Control Point*: Any point, step, or procedure at which biological, physical, or chemical factors can be controlled.

1.5 *Corrective Action*: Procedures to be followed when a deviation occurs.

1.6 *Criterion*: A requirement on which a judgement or decision can be based.

1.7 *Critical Control Point (CCP)*: A point, step, or procedure at which control can be applied and a food safety hazard can be prevented, eliminated, or reduced to acceptable levels.

1.8 *Critical Defect*: A deviation at a CCP which may result in a hazard.

1.9 *Critical Limit*: A criterion that must be met for each preventive measure associated with a critical control point.

1.10 *Deviation*: Failure to meet a critical limit.

1.11 *HACCP Plan*: The written document which is based upon the principles of HACCP and which delineates the procedures to be followed to assure the control of a specific process or procedure.

1.12 *HACCP System*: The result of the implementation of the HACCP plan.

1.13 *HACCP Team*: The group of people who are responsible for developing a HACCP plan.

1.14 *HACCP Plan Revalidation*: One aspect of verification in which a documented periodic view of the HACCP plan is done by the HACCP team with the purpose of modifying the HACCP plan as necessary.

1.15 *HACCP Plan Validation*: The initial review by the HACCP team to ensure that all elements of the HACCP plan are accurate.

1.16 *Hazard*: A biological, chemical, or physical property that may cause a food to be unsafe for consumption.

1.17 *Monitor*: To conduct a planned sequence of observations or measurements to assess whether a CCP is under control and to produce an accurate record for future use in verification.

1.18 *Preventive Measure*: Physical, chemical, or other factors that can be used to control an identified health hazard.

1.19 *Random Checks*: Observations or measurements which are performed to supplement the scheduled evaluations required by the HACCP plan.

1.20 *Risk*: An estimate of the likely occurrence of a hazard.

1.21 *Sensitive Ingredient*: An ingredient known to have been associated with a hazard and for which there is reason for concern.

1.22 *Severity*: The seriousness of a hazard.

1.23 *Target Levels*: Criteria which are more stringent than critical limits and which are used by an operator to reduce the risk of a deviation.

1.24 *Verification*: The use of methods, procedures, or tests in addition to those used in monitoring to determine if the HACCP system is in compliance with the HACCP plan and/or whether the HACCP plan needs modification and revalidation.

2.0 PURPOSE AND PRINCIPLES

HACCP is a systematic approach to food safety consisting of seven principles:

2.1 Conduct a hazard analysis. Prepare a list of steps in the process where significant hazards occur and describe the preventive measures.

2.2 Identify the CCPs in the process.

2.3 Establish critical limits for preventive measures associated with each identified CCP.

2.4 Establish CCP monitoring requirements. Establish procedures for using the results of monitoring to adjust the process and maintain control.

2.5 Establish corrective actions to be taken when monitoring indicates that there is a deviation from an established critical limit.

2.6 Establish effective record-keeping procedures that document the HACCP system.

2.7 Establish procedures for verification that the HACCP system is working correctly.

3.0 EXPLANATION AND APPLICATION OF PRINCIPLES

The HACCP concept is relevant to all stages throughout the food chain from growing, harvesting, processing, manufacturing, distributing, and merchandising to preparing food for consumption. Certain points in the food chain are better suited to the application of the HACCP principles. For example, food manufacturing facilities are very well suited to the adoption of the HACCP concept. The Committee recommends the adoption of HACCP to the fullest extent possible and reasonable throughout the food chain.

The Committee recognizes that education and training is an important element of the HACCP concept. Employees who will be responsible for the HACCP program must be adequately trained in the principles of HACCP, its application and implementation. However, education and training programs do not have to be limited to those directly involved with HACCP and its implementation. Educational and training programs should be designed to address the needs of industry, government and academic personnel, as well as consumers. Educating home food handlers in the recognition and application of critical control points will improve the safety of foods prepared in the home. It is recommended that educational materials be provided to home food handlers that address the safe acquisition and proper handling of foods.

The following figure (C-1) lists steps used in the application of Principle 1.

3.1 *Assemble the HACCP team.*

The first step in developing a HACCP plan is to assemble a HACCP team consisting of individuals who have specific knowledge and expertise appropriate to the product and process. It is the team's responsibility to develop each step of the HACCP plan. The team should be multidisciplinary (e.g., engineering, production, sanitation, quality assurance, food microbiology). The team should include local personnel who are directly involved in the daily processing activities as they are more familiar with the variability and limitations of the operation. In addition, this fosters a sense of ownership among those who must implement the plan. The HACCP team might require outside experts who are knowledgeable in the potential microbiological and other public health risks associated with the product and the process. However, a plan which is developed totally by outside sources will likely be erroneous, incomplete, and lacking in support at the local level.

Due to the technical nature of the information required for a hazard analysis, it is recommended that experts who are knowledgeable about the food and process should either participate in or verify the completeness of the hazard analysis and the HACCP plan. These individuals should have the knowledge and experience to correctly (a) identify potential hazards; (b) assign levels of severity and risk; (c) recommend controls, criteria, and procedures for monitoring and verification; (d) recommend appropriate corrective actions when a deviation occurs; (e) recommend research related to the HACCP plan if important information is not known; and (f) predict the success of the HACCP plan.

3.2 *Describe the food and the method of its distribution.*

A separate HACCP plan must be developed for each food product that is being processed in the establishment. The HACCP team must first fully describe the food. This consists of a full description of the food including the recipe or formulation. The method of distribution should be described along with information on whether the food is to be distributed frozen, refrigerated, or shelf stable. Consideration should also be given to the potential for abuse in the distribution channel and by consumers.

3.3 *Identify the intended use and consumers of the food.*

The intended use of the food should be based upon the normal use of the food by end users or consumers.
The intended consumers may be the general public or a particular segment of the population, such as infants, the elderly, etc.

3.4 *Develop a flow diagram which describes the process.*

The purpose of the diagram is to provide a clear, simple description of the steps involved in the process. The diagram will be helpful to the HACCP team in its subsequent work. The diagram can also serve as a future guide for others (e.g., regulatory officials and customers) who must understand the process for their verification activities.

The scope of the flow diagram must cover all the steps in the process which are directly under the control of the establishment. In addition, the flow diagram can include steps in the food chain which are before and after the processing that occurs in the establishment. For the sake of simplicity, the flow diagram should consist solely of words, not engineering drawings.

3.5 *Verify flow diagram.*

The HACCP team should inspect the operation to verify the accuracy and completeness of the flow diagram. The diagram should be modified as necessary.

3.6 Principle No. 1: *Conduct a hazard analysis. Prepare a list of steps in the process where significant hazards occur and describe the preventive measures.*

The HACCP team next conducts a hazard analysis and identifies the steps in the process where hazards of potential significance can occur. For inclusion in the list, the hazards must be of such a nature that their prevention, elimination or reduction to acceptable levels is essential to the production of a safe food. Hazards which are of a low risk and not likely to occur would not require further consideration. The team must then consider what preventive measures, if any, exist which can be applied for each hazard. Preventive measures are physical, chemical, or other factors that can be used to control an identified health hazard. More than one preventive measure may be required to control a specific hazard. More than one hazard may be controlled by a specified preventive measure.

The hazard analysis and identification of associated preventive measures accomplishes three purposes: First, those hazards of significance and associated preventive measures are identified. Second, the analysis can be used to modify a process or product to further assure or improve safety. Third, the analysis provides a basis for determining CCPs in Principle 2 (Section 3.7).

The hazard analysis procedure differs from the Committee's original HACCP document. This does not negate the validity of current plans based on the earlier method of hazard analysis. The procedures outlined in this document are recommended for future use.

The hazard analysis consists of asking a series of questions which are appropriate to the specific food process and establishment. It is not possible in these recommendations to provide a list of all the questions which may be pertinent to a specific food or process. The hazard analysis should question the effect of a variety of factors upon the safety of the food. Table C-1 lists examples of questions that may be considered during the hazard analysis. The original hazard analysis format is included as a Table C-2 for comparison.

The hazard analysis must consider factors which may be beyond the immediate control of the processor. For example, product distribution may be beyond the immediate control of the processor, but information on how the food will be distributed could influence, for example, how the food will be processed.

During the hazard analysis, the potential significance of each hazard should be assessed by considering its risk and severity. Risk is an estimate of the likely occurrence of a hazard. The estimate of risk is usually based upon a combination of experience, epidemiological data, and information in the technical literature. Severity is the seriousness of a hazard.

The HACCP team has the initial responsibility to decide which hazards are significant and must be addressed in the HACCP plan. This decision can be debatable. There may be differences of opinion, even among experts, as to the risk of a hazard. The HACCP team must rely upon the opinion of the experts who assist in the development of the HACCP plan.

During the hazard analysis, safety concerns must be differentiated from quality concerns. Hazard is defined as a biological, chemical or physical property that may cause a food

TABLE C-1   Examples of Questions to be Considered in a Hazard Analysis

The hazard analysis consists of asking a series of questions which are appropriate to each step in a HACCP plan. It is not possible in these recommendations to provide a list of all the questions which may be pertinent to a specific food or process. The hazard analysis should question the effect of a variety of factors upon the safety of the food.

A. Ingredients
    1. Does the food contain any sensitive ingredients that may present microbiological hazards (e.g., *Salmonella, Staphylococcus aureus*); chemical hazards (e.g., aflatoxin, antibiotic or pesticide residues); or physical hazards (stones, glass, metal)?
    2. Is potable water used in formulating or in handling the food?
B. Intrinsic Factors
Physical characteristics and composition (e.g., pH, type of acidulents, fermentable carbohydrate, water activity, preservatives) of the food during and after processing.
    1. Which intrinsic factors of the food must be controlled in order to assure food safety?
    2. Does the food permit survival or multiplication of pathogens and/or toxin formation in the food during processing?
    3. Will the food permit survival or multiplication of pathogens and/or toxin formation during subsequent steps in the food chain?
    4. Are there other similar products in the market place? What has been the safety record for these products?
C. Procedures used for processing
    1. Does the process include a controllable processing step that destroys pathogens? Consider both vegetative cells and spores.
    2. Is the product subject to recontamination between processing (e.g., cooking, pasteurizing) and packaging?
D. Microbial content of the food
    1. Is the food commerically sterile (e.g., low acid canned food)?
    2. Is it likely that the food will contain viable sporeforming or nonsporeforming pathogens?
    3. What is the normal microbial content of the food?
    4. Does the microbial population change during the normal time the food is stored prior to consumption?
    5. Does the subsequent change in microbial population alter the safety of the food, pro or con?
E. Facility design
    1. Does the layout of the facility provide an adequate separation of raw materials from ready-to-eat foods if this is important to food safety?
    2. Is positive air pressure maintained in product packaging areas? Is this essential for product safety?
    3. Is the traffic pattern for people and moving equipment a significant source of contamination?
F. Equipment design
    1. Will the equipment provide the time-temperature control that is necessary for safe food?
    2. Is the equipment properly sized for the volume of food that will be processed?
    3. Can the equipment be sufficiently controlled so that the variation in performance will be within the tolerances required to produce a safe food?

(continued)

TABLE C-1   Continued

4. Is the equipment reliable or is it prone to frequent breakdowns?
5. Is the equipment designed so that it can be cleaned and sanitized?
6. Is there a chance for product contamination with hazardous substances; e.g., glass?
7. What product safety devices are used to enhance consumer safety?
   - metal detectors
   - magnets
   - sifters
   - filters
   - screens
   - thermometers
   - deboners
   - dud detectors

G. Packaging
1. Does the method of packaging affect the multiplication of microbial pathogens and/or the formation of toxins?
2. Is the package clearly labeled "Keep Refrigerated" if this is required for safety?
3. Does the package include instructions for the safe handling and preparation of the food by the end user?
4. Is the packaging material resistant to damage thereby preventing the entrance of microbial contamination?
5. Are tamper-evident packaging features used?
6. Is each package and case legibly and accurately coded?
7. Does each package contain the proper label?

H. Sanitation
1. Can sanitation impact upon the safety of the food that is being processed?
2. Can the facility and equipment be cleaned and sanitized to permit the safe handling of food?
3. Is it possible to provide sanitary conditions consistently and adequately to assure safe foods?

I. Employee health, hygiene and education
1. Can employee health or personal hygiene practices impact upon the safety of the food being processed?
2. Do the employees understand the process and the factors they must control to assure the preparation of safe foods?
3. Will the employees inform management of a problem which could impact upon safety of the food?

J. Conditions of storage between packaging and the end user
1. What is the likelihood that the food will be improperly stored at the wrong temperature?
2. Would an error in improper storage lead to a microbiologically unsafe food?

K. Intended use
1. Will the food be heated by the consumer?
2. Will there likely be leftovers?

L. Intended consumer
1. Is the food intended for the general public?
2. Is the food intended for consumption by a population with increased susceptibility to illness (e.g., infants, the aged, the infirmed, immunocompromised individuals)?

4.1.2 Hazard Analysis and Assignment of Risk Categories

4.1.2.1 Rank the food according to hazard characteristics A through F, using a plus (+) to indicate a potential hazard. The number of pluses will determine the risk category. A model diagram outlining this concept is given under section 4.1.3. As indicated, if the product falls under Hazard Class A, it should automatically be considered Risk Category VI.

Hazard A: A special class that applies to nonsterile products designated and intended for consumption by at-risk populations, e.g., infants, the aged, the infirm, or immuno-compromised individuals.

Hazard B: The product contains "sensitive ingredients" in terms of microbiological hazards.

Hazard C: The process does not contain a controlled processing step that effectively destroys harmful microogranisms.

Hazard D: The product is subject to recontamination after processing before packaging.

Hazard E: There is substantial potential for abusive handling in distribution or in consumer handling that could render the product harmful when consumed.

Hazard F: There is no terminal heat process after packaging or when cooked in the home.

Note: Hazards can also be stated for chemical or physical hazards, particularly if a food is subject to them.

4.1.2.2 Assignment of risk category (based on ranking by hazard chaacteristics):

Category VI. A special category that applies to nonsterile products designated and intended for consumption by at-risk populations, e.g., infants, the aged, the infirm, or immunocompromised individuals. All six hazard characteristics must be considered.

Category V. Food products subject to all five general hazard characteristics.
    Hazard Class B, C, D, E. F

Category IV. Food products subject to four general hazard characteristics.

Category III. Food products subject to three of the general hazard characteristics.

Category II. Food products subject to two of the general hazard characteristics.

Category I. Food products subject to one of the general hazard characteristics.

Category 0. Hazard Class-No hazard.

Note: Ingredients are treated in the same manner in respect to how they are received at the plant, *before* processing. This permits determination of how to reduce risk in the food system.

4.1.3 It is recommended that a chart be utilized that provides assessment of a food by hazard characteristic and risk category. A format for this chart is given as follows:

Food Ingredient or Product	Hazard Characteristics (A,B,C,D,E,F)	Risk Category (VI,V,IV,III,II,I,0)
T	A + (Special Category)*	VI
U	Five + 's (B through F)	V
V	Four + 's (B through F)	IV
W	Three + 's (B through F)	III
X	Two + 's (B through F)	I
Y	One + (B through F)	I
Z	No + 's	0

*Hazard characteristics A automatically is Risk Category VI, but any combination of B through F may also be present.

192

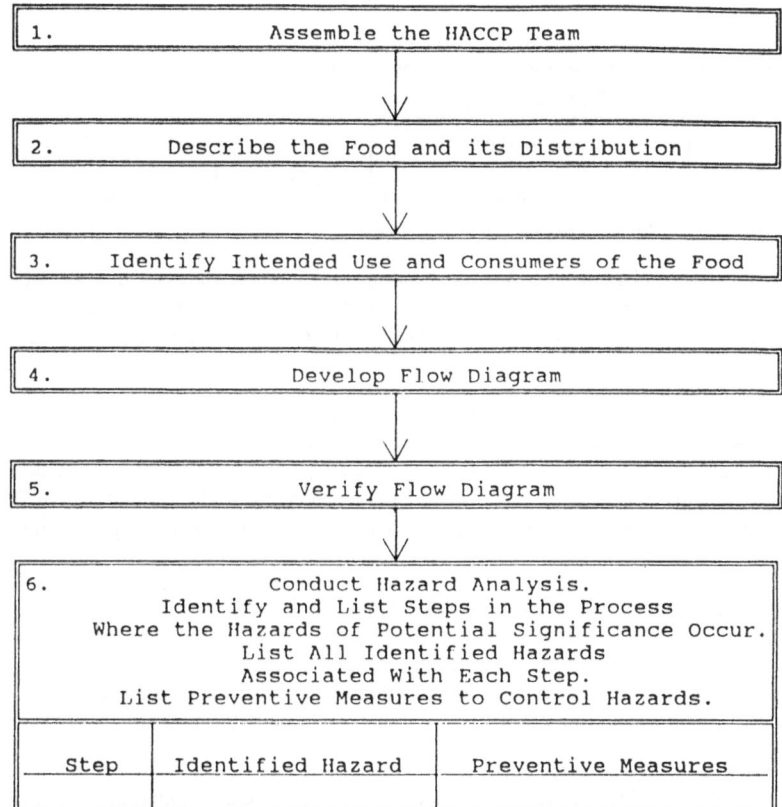

FIGURE C-1   First six steps for the development of a HACCP Plan.

to be unsafe for consumption. The term hazard as used in this document is limited to safety. The HACCP team must make the determination whether a potential problem is a safety concern and of its likelihood of occurrence.

Upon completion of the hazard analysis, the significant hazards associated with each step in the flow diagram should be listed along with any preventive measures to control the hazards (step 6 of Fig. C-1). This tabulation will be used in Principle 2 to determine CCPs.

For example, if a HACCP team were to conduct a hazard analysis for the production of frozen cooked beef patties (Table C-3), enteric pathogens in the raw meat would be identified as a potential hazard. Cooking is a preventive measures which can be used to eliminate this hazard. Thus, cooking would be listed along with the hazard (i.e., enteric pathogens) and the preventive measure as follows:

Step	Identified Hazard	Preventive Measures
5. Cooking	Enteric pathogens	Cooking sufficiently to kill enteric pathogens

3.7 Principle No. 2: *Identify the CCPs in the process.*

3.7.1 A critical control point is defined as a point, step or procedure at which control can be applied and a food safety hazard can be prevented, eliminated, or reduced to acceptable levels. All significant hazards identified by the HACCP team during the hazard analysis must be addressed.

The information developed during the hazard analysis in section 3.6 should enable the HACCP team to identify which steps in the process are CCPs. Identification of each CCP can be facilitated by the use of a CCP decision tree (Fig. C-2). All hazards which reasonably could be expected to occur should be considered. Application of the CCP decision tree can help determine if a particular step is a CCP for a previously identified hazard.

Critical control points are located at any point where hazards need to be either prevented, eliminated, or reduced to acceptable levels. For example, a specified heat process, at a given time and temperature to destroy a specified microbiological pathogen, is a CCP. Likewise, refrigeration required to prevent hazardous microorganisms from multiplying, or the adjustment of a food to a pH necessary to prevent toxin formation are also CCPs.

Examples of CCPs may include, but are not limited to: cooking, chilling, specific sanitation procedures, product formulation control, prevention of cross contamination, and certain aspects of employee and environmental hygiene.

CCPs must be carefully developed and documented. In addition, they must be used only for purposes of product safety.

Different facilities preparing the same food can differ in the risk of hazards and the points, steps, or procedures which are CCPs. This can be due to differences in each facility such as layout, equipment, selection of ingredients, or the process that is employed. Generic HACCP plans can serve as useful guides; however, it is essential that the unique conditions within each facility be considered during the development of a HACCP plan.

In addition to CCPs, non food safety concerns may be addressed at control points. These control points will not be further discussed in this document because they do not relate to food safety and are not included in the HACCP plan.

3.8 Principle No. 3: *Establish critical limits for preventive measures associated with each identified CCP.*

3.8.1 A critical limit is defined as a criterion that must be met for each preventive measure

TABLE C-3   Example of a Flow Diagram for the Production of Frozen Cooked Beef Patties

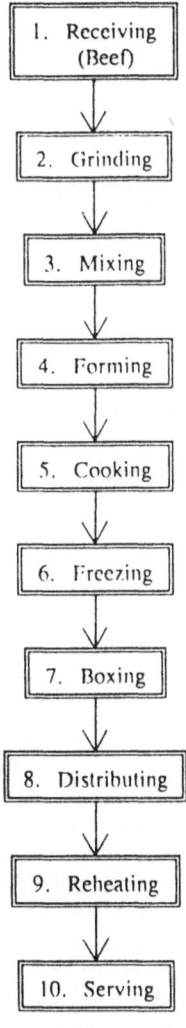

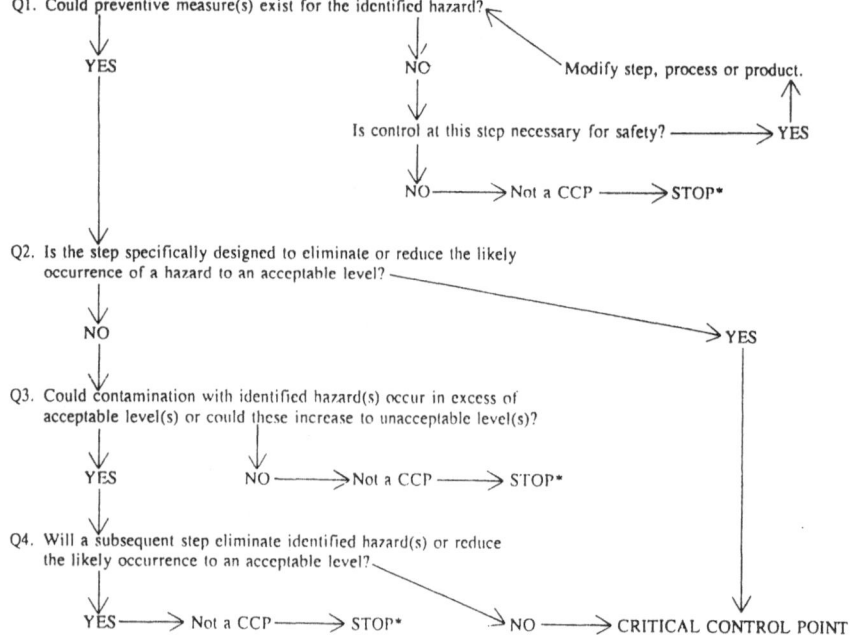

Q1. Could preventive measure(s) exist for the identified hazard?

YES

NO

Modify step, process or product.

Is control at this step necessary for safety? ⟶ YES

NO ⟶ Not a CCP ⟶ STOP*

Q2. Is the step specifically designed to eliminate or reduce the likely occurrence of a hazard to an acceptable level?

NO

YES

Q3. Could contamination with identified hazard(s) occur in excess of acceptable level(s) or could these increase to unacceptable level(s)?

YES

NO ⟶ Not a CCP ⟶ STOP*

Q4. Will a subsequent step eliminate identified hazard(s) or reduce the likely occurrence to an acceptable level?

YES ⟶ Not a CCP ⟶ STOP*

NO ⟶ CRITICAL CONTROL POINT

*Proceed to next step in the described process

FIGURE C-2   CCP decision tree.

associated with a CCP. Each CCP will have one or more preventive measures that must be properly controlled to assure prevention, elimination or reduction of hazards to acceptable levels. Each preventive measure has associated with it, critical limits that serve as boundaries of safety for each CCP. Critical limits may be set for preventive measures such as temperature, time, physical dimensions, humidity, moisture level, water activity (a_w), pH, titratable acidity, salt concentration, available chlorine, viscosity, preservatives, or sensory information such as texture, aroma, and visual appearance. Critical limits may be derived from sources such as regulatory standards and guidelines, literature surveys, experimental studies, and experts. The food industry is responsible for engaging competent authorities to validate that the critical limits will control the identified hazard.

For example, an acidified beverage that requires only hot fill and hold as a thermal process may have acid addition as a CCP. If insufficient acid is added, the product would be underprocessed and allow the growth of pathogenic sporeforming bacteria. One preventive measure for this CCP may be pH with a critical limit of pH 4.6. The critical limit for controlling a potential health hazard may be different from criteria associated with quality factors. For example, the product may be of unacceptable quality when the pH exceeds 3.8; however, a health hazard is avoided when the critical limit of pH 4.6 is not exceeded.

In some cases, processing variations may require certain target levels to assure that critical limits are attained. For example, a preventive measure and critical limit may be an internal product temperature of 160°F (71.1°C) during one stage of a process. The oven temperature, however, may be ±5°F (2.8°C) at 160°F; thus an oven target temperature would have to be greater than 165°F (73.9°C) so that no product receives a cook of less than 160°F.

An example for Principle 3 is the cooking of beef patties (Table C-3). The process should be designed to eliminate the most heat-resistant vegetative pathogen which could reasonably be expected to be in the product. Criteria may be required for factors such as temperature, time and meat patty thickness. Technical development of the appropriate critical limits requires accurate information on the probable maximum numbers of these microorganisms in the meat and their heat resistance. The relationship between the CCP and its critical limits for the meat patty example is shown below:

Process Step	CCP	Critical Limits
5. Cooking	YES	Minimum internal temperature of patty: e.g. 145°F Oven temperature: _____°F Time; rate of heating and cooling (belt speed in rpm): _____ rpm Patty thickness: _____ in. Patty composition: e.g. all beef Oven humidity: _____ % RH

3.9 Principle No. 4: *Establish CCP monitoring requirements. Establish procedures for using the results of monitoring to adjust the process and maintain control.*

3.9.1 Monitoring is a planned sequence of observations or measurements to assess whether a CCP is under control and produce an accurate record (Table C-4) for future use in verification. Monitoring serves three main purposes. First, monitoring is essential to food safety management in that it tracks the system's operation. If monitoring indicates that there is a trend towards loss of control, i.e., exceeding a target level, then action can be taken to bring the process back into control before a deviation occurs. Second, monitoring is used to determine when there is loss of control and a deviation occurs at a CCP, i.e., exceeding the critical limit. Corrective action then must be taken. Third, it provides written documentation for use in verification of the HACCP plan.

An unsafe food may result if a process is not properly controlled and a deviation occurs. Because of the potentially serious consequences of a critical defect, monitoring procedures must be effective. Ideally, monitoring should be at the 100% level. Continuous monitoring is possible with many types of physical and chemical methods. For example, the temperature and time for the scheduled thermal process of low-acid canned foods is recorded continuously on temperature recording charts. If the temperature falls below the scheduled temperature or the time is insufficient, as recorded on the chart, the retort load is retained

TABLE C-4   Examples of HACCP Records

A. Ingredients
   1. Supplier certification documenting compliance with processor's specifications.
   2. Processor audit records verifying supplier compliance.
   3. Storage temperature record for temperature sensitive ingredients.
   4. Storage time records of limited shelf life ingredients.
B. Records relating to product safety
   1. Sufficient data and records to establish the efficacy of barriers in maintaining product safety.
   2. Sufficient data and records establishing the safe shelf life of the product; if age of product can affect safety.
   3. Documentation of the adequacy of the processing procedures from a knowledgeable process authority.
C. Processing
   1. Records from all monitored CCPS.
   2. Records verifying the continued adequacy of the processes.
D. Packaging
   1. Records indicating compliance with specifications of packaging materials.
   2. Records indicating compliance with sealing specifications.
E. Storage and distribution
   1. Temperature records.
   2. Records showing no product shipped after shelf life date on temperature sensitive products.
F. Deviation and corrective action records
G. Validation records and modification to the HACCP plan indicating approved revisions and changes in ingredients, formulations, processing, packing and distribution control, as needed.
H. Employee training records

as a process deviation. Likewise, pH measurement may be performed continually in fluids or by testing of a batch before processing. There are many ways to monitor CCP limits on a continuous or batch basis and record the data on charts. Continuous monitoring is always preferred when feasible. Equipment must be carefully calibrated for accuracy.

Assignment of the responsibility for monitoring is an important consideration for each CCP. Specific assignments will depend on the number of CCPs and preventive measures and the complexity of monitoring. Such individuals are often associated with production (e.g., line supervisors, selected line workers and maintenance personnel) and, as required, quality control personnel. Those individuals monitoring CCPs must be trained in the technique used to monitor each preventive measure; fully understand the purpose and importance of monitoring; have ready access to the monitoring activity; be unbiased in monitoring and reporting; and accurately report the monitoring activity. Personnel assigned the monitoring activity must report the results. Unusual occurrences must be reported immediately so that adjustments can be made in a timely manner to assure that the process remains under control. The person responsible for monitoring must also report a process or product that does not meet critical limits so that immediate corrective action can be taken.

When it is not possible to monitor a critical limit on a continuous basis, it is necessary to establish that the monitoring interval will be reliable enough to indicate that the hazard is under control. Statistically designed data collection or sampling systems lend themselves to this purpose. When using statistical process control, it is important to recognize that critical limits must not be exceeded. For example, when pH of 4.6 or less is required for product safety, the maximum pH of the product may be set at a target that is below pH 4.6 to compensate for variation.

Most monitoring procedures for CCPs will need to be done rapidly because they relate to on-line processes and there will not be time for lengthy analytical testing. Microbiological testing is seldom effective for monitoring CCPs due to their time-consuming nature. Therefore, physical and chemical measurements are preferred because they may be done rapidly and can indicate the conditions of microbiological control in the process.

Examples of measurements for monitoring include:

Visual observations
Temperature
Time
pH
Moisture level

Random checks may be useful for supplementing the monitoring of certain CCPs. They may be used to check incoming pre-certified ingredients, assess equipment and environmental sanitation, airborne contamination, cleaning and sanitizing of gloves and any place where follow-up is needed. Random checks may consist of physical and chemical testing and, as appropriate, microbiological tests.

With certain foods, microbiologically sensitive ingredients, or imports, there may be no alternative to microbiological testing. However, it is important to recognize that a sampling frequency that is adequate for reliable detection of low levels of pathogens is seldom possible because of the large number of samples needed. For this reason, microbiological testing has limitations in a HACCP system, but is valuable as a means of establishing and randomly verifying the effectiveness of control at CCPs (challenge tests, random testing or for troubleshooting).

All records and documents associated with CCP monitoring must be signed or initialled by the person doing the monitoring.

3.10 Principle No. 5: *Establish corrective action to be taken when monitoring indicates that there is a deviation from an established critical limit.*

3.10.1 The HACCP system for food safety management is designed to identify potential health hazards and to establish strategies to prevent their occurrence. However, ideal circumstances do not always prevail and deviations from established processes may occur. For instances where there is a deviation from established critical limits, corrective action plans must be in place to (a) determine the disposition of non-compliance product, (b) fix or correct the cause of non-compliance to assure that the CCP is under control, and (c) maintain records of the corrective actions that have been taken where there has been a deviation from critical limits. Because of the variations in CCPs for different foods and the diversity of possible deviations, specific corrective action plans must be developed for each CCP. The actions must demonstrate the CCP has been brought under control. Individuals who have a thorough understanding of the process, product and HACCP plan are to be assigned responsibility for taking corrective action. Corrective action procedures must be documented in the HACCP plan.

Should a deviation occur, the plant will place the product on hold pending completion of appropriate corrective actions and analyses. As appropriate, scientific experts and regulatory agencies are to be consulted to determine additional testing and disposition of the product.

Identification of deviant lots and corrective actions taken to assure safety of these lots must be noted in the HACCP record and remain on file for a reasonable period after the expiration date or expected shelf life of the product.

3.11 Principle No. 6: *Establish effective recordkeeping procedures that document the HACCP system.*

3.11.1 The approved HACCP plan and associated records must be on file at the food establishment. Generally, the records utilized in the total HACCP system will include the following:

1. The HACCP plan

Listing of the HACCP team and assigned responsibilities.
Description of the product and its intended use.
Flow diagram for the entire manufacturing process indicating CCPs.
Hazards associated with each CCP and preventive measures.
Critical limits
Monitoring system.
Corrective action plans for deviations from critical limits.
Recordkeeping procedures.
Procedures for verification of HACCP system.

In addition to listing the HACCP team, product description and uses, and providing a flow diagram, other information in the HACCP plan can be tabulated as follows:

Process Step	CCP	Chem. Phys. Biolog. Hazards	Critical Limits	Monitoring Procedures/ Frequency/ Person(s) Responsible	Corrective Action(s)/ Person(s) Responsible	HACCP Records	Verification Procedure/Person(s) Responsible
1.	Yes or No	1. 2. 3. etc.					

2. Records obtained during the operation of the plan. (Table C-4)

3.12 Principle No. 7: *Establish procedures for verification that the HACCP system is working correctly.*

The National Academy of Sciences (1985)[1] pointed out that the major infusion of science in a HACCP system centers on proper identification of the hazards, critical control points, critical limits, and instituting proper verification procedures. These processes should take place during the development of the HACCP plan. There are four processes involved in verification.

3.12.1 The first is the scientific or technical process to verify that critical limits at CCPs are satisfactory. This process is complex and requires intensive involvement of highly skilled professionals from a variety of disciplines capable of doing focused studies and analyses. The process consists of a review of the critical limits to verify that the limits are adequate to control the hazards that are likely to occur.

3.12.2 The second process of verification ensures that the facility's HACCP plan is functioning effectively. A functioning HACCP system requires little end-product sampling, since appropriate safeguards are built in early in the process. Therefore, rather than relying on end-product sampling, firms must rely on frequent reviews of their HACCP plan, verification that the HACCP plan is being correctly followed, review of CCP records, and determinations that appropriate risk management decisions and product dispositions are made when process deviations occur.

3.12.3 The third process consists of documented periodic revalidations, independent of audits or other verification procedures, that must be performed to ensure the accuracy of the HACCP plan. Revalidations are performed by a HACCP team on a regular basis and/or whenever significant product, process or packaging changes require modification of the HACCP plan. The revalidation includes a documented on-site review and verification of all flow diagrams and CCPs in the HACCP plan. The HACCP team modifies the HACCP plan as necessary.

3.12.4 The fourth process of verification deals with the government's regulatory responsibility and actions to ensure that the establishment's HACCP system is functioning satisfactorily.

Examples of verification activities are included as Table C-5.

TABLE C-5   Examples of Verification Activities

A. Verification procedures may include:
    1. Establishment of appropriate verification inspection schedules.
    2. Review of the HACCP plan.
    3. Review of CCP records.

(continued)

---

[1]*An Evaluation of the Role of Microbiological Criteria for Foods and Food Ingredients.* National Academy of Sciences, National Academy Press, Washington DC.

TABLE C-5   Continued

4. Review of deviations and dispositions.
5. Visual inspections of operations to observe if CCP are under control.
6. Random sample collection and analysis.
7. Review of critical limits to verify that they are adequate to control hazards.
8. Review of written record of verification inspections which certifies compliance with the HACCP plan or deviations from the plan and the corrective actions taken.
9. Validation of HACCP plan, including on-site review and verification of flow diagrams and CCPs.
10. Review of modifications of the HACCP plan.
B. Verification inspections should be conducted:
    1. Routinely, or on an unannounced basis, to assure selected CCP are under control.
    2. When it is determined that intensive coverage of a specific commodity is needed because of new information concerning food safety.
    3. When foods produced have been implicated as a vehicle of foodborne disease.
    4. When requested on a consultative basis or established criteria have not been met.
    5. To verify that changes have been implemented correctly after a HACCP plan has been modified.
C. Verification reports should include information about:
    1. Existence of a HACCP plan and the person(s) responsible for administering and updating the HACCP plan.
    2. The status of records associated with CCP monitoring.
    3. Direct monitoring data of the CCP while in operation.
    4. Certification that monitoring equipment is properly calibrated and in working order.
    5. Deviations and corrective actions.
    6. Any samples analyzed to verify that CCP are under control. Analyses may involve physical, chemical, microbiological or organoleptic methods.
    7. Modifications to the HACCP plan.
    8. Training and knowledge of individuals responsible for monitoring CCPs.

Index